DEFINING AEROSPACE POLICY

Defining Aerospace Policy

Essays in Honor of Francis T. Hoban

Edited by

KENNETH BUTTON
JULIANNE LAMMERSEN-BAUM
ROGER STOUGH
George Mason University, USA

Routledge
Taylor & Francis Group
LONDON AND NEW YORK

First published 2004 by Ashgate Publishing

2 Park Square, Milton Park, Abingdon, Oxfordshire OX14 4RN
711 Third Avenue, New York, NY 10017

Routledge is an imprint of the Taylor & Francis Group, an informa business

First issued in paperback 2018

British Library Cataloguing in Publication Data
Defining aerospace policy : essays in honor of Francis T. Hoban
1. Astronautics and state - United states 2. Astronautics
3. Navigation (Aeronautics)
I. Button, Kenneth John II. Lammersen-Baum, Julianne
III. Stough, Roger IV. Hoban, Francis T.
629.4'0973

Library of Congress Cataloging-in-Publication Data
Defining aerospace policy : essays in honor of Francis T. Hoban / edited by Kenneth Button, Julianne Lammersen-Baum, and Roger Stough.
p. cm.
Includes bibliographical references and index.
ISBN 0-7546-4225-9
1. Aerospace industries--Government policy. I. Hoban, Francis T. II. Button, Kenneth John. III. Lammersen-Baum, Julianne. IV. Stough, Roger.

HD9711.5.A2D436 2004
387.8--dc22

2004041040

Transfered to Digital Printing in 2011

ISBN 978-0-7546-4225-1 (hbk)
ISBN 978-1-138-38113-1 (pbk)

Contents

Preface

This collection of chapters has been prepared as a tribute to the work of Frank Hoban. Frank died suddenly in 2002 after spending most of his working life involved in the aerospace industry. After his long career at the National Aeronautics and Space Administration, he moved to the School of Public Policy where he continued his association with the aerospace community by heading up the NASA/George Mason University Continuing Career Program.

The chapters here, some of which Frank worked on, stem mainly from the later years of his career and are largely contributed by those who worked with him at George Mason University. They are not in any way meant to reflect all of Frank's interests, nor his contributions to aerospace administration and research. Rather they just touch upon some of the issues that interested Frank. We hope they provide a small tribute to his contribution to the aerospace world.

The editors of the book would like to acknowledge that it is published under a cooperative agreement (NCC5-360) with the National Aeronautics and Space Administration.

Kenneth Button, Julianne Lammersen-Baum and Roger Stough

Authors' Profiles

Kenneth Button BA (East Anglia), MA (Leeds), PhD (Loughborough), FILT, FIHT, FCITI is Professor of Public Policy, Director of the Center for Transportation Policy and Operations and Director of the Center for Aerospace Policy in the School of Public Policy, George Mason University. Dr. Button's academic training was in the fields of economics, econometrics and transportation planning. He has published, or has in press, some 80 books, over 400 papers in leading academic journals, and forecasting software. He has also given written and oral evidence to the US Congressional Transportation Committee and to both the UK House of Lords and UK House of Commons Transport Committees. He is editor of the academic journals *Transportation Research D: Transport and the Environment* and of the *Journal of Air Transport Management* and is on the editorial boards of nine other journals. He is on the scientific committee of the World Conference on Transport Research.

Kingsley Haynes BA (Western Michigan University), MA (Rutgers University) PhD (Johns Hopkins University) is Dean of the School of Public Policy, George Mason University. He was a founding faculty member at the Lyndon B. Johnson School of Public affairs at the University of Texas, Austin, Chair of the Department of Geography at Boston University and Chair of the Urban, Regional Analysis and Planning Faculty at Indiana University. His research work is in the field of regional economic development policy with over 200 articles and 10 books to his name. He has been an active participant in economic development activities in Texas, the US Midwest and in Malaysia, Brazil, Southeast Asia and the Middle East where he was awarded prizes for his work on the Nile and building regional development programs in the nations of the Pacific Rim. In 1997 he was recognized with the Boyce Award for his service as the 1995-97 President of the 70 nation Regional Science Association International. He was awarded the Anderson Medal for work in applied research in 2000, was elected to the National Academy of Public Administration in 2002, and also in 2002 was awarded the Ullman Prize for his research contributions in transportation.

Clay R. Hicks began his aerospace career as a co-op student for NASA's Langley Research Center in the late 1950s. Following his graduation from Virginia Polytechnic, the University of Virginia with an engineering degree, he was assigned to the Space Task Group that was planning the Mercury Project. He later moved to Houston Texas with the rest of the task group. There he was assigned to the Mission Operations Directorate. In the 1980s, he transferred to the Office of Space Flight at NASA Headquarters. He became one of the initial members of the

Space Station Task Force. He left NASA to work for the Rockwell Corporation as a Washington technical representative and later performed the same function for the Boeing Company. Clay Hicks is now an aerospace consultant.

Francis T. Hoban BA (St. Louis University) MA (George Washington University) MBA (College of William and Mary). He spent more than forty years in the aerospace industry and has written extensively about project management. In 1970 he was selected to serve on the staff of the NASA Administrator as an Assistant Executive Secretary. In 1973 he joined the Low Cost Systems Office as the Director of Business Practices. In 1976 he was selected as a Presidential exchange Executive and spent a year with the General Motors Corporation. Returning to NASA, Mr. Hoban was assigned to implement the Civil Service Reform Act. Later, He was asked to join the President's Commission on the Accident at Three Mile Island as its Director of Administration. In 1982, Mr. Hoban joined the Space Station Task Force. In 1986, after the *Challenger* accident, he led a program to train the NASA workforce engaged in managing programs and projects and this earned him NASA's Exceptional Achievement Medal. From 1994 he was an Adjunct Professor in George Mason University's School of Public Policy and Director of the NASA/GMU Continuing Career Program.

Lawrence S. Jessie BS (Brooklyn College), MA (Columbia University). He was at the time of his retirement at NASA's Goddard Space Flight Center where he managed a satellite/laser project. His earlier government career included assignments as a physicist with the US Army, the National Oceanic and Atmospheric Administration, and the Defense Intelligence Agency. He also spent seven years at NASA Headquarters as a technical manager in the Office of Policy Coordination and International Relations. For the last 25 years he has been an adjunct professor in astronomy and space science at Prince George's Community College.

Julianne Lammersen-Baum BA (Loyola Marymount University) MA (Georgetown University). She initially worked at the International Bank for Reconstruction and Development as an economic analyst dealing with international capital markets. She has extensive experience in consultancy work in the field of economic analysis. Recently she has been involved with the George Mason University NASA Continuing Career Program and has been responsible for production of the final report on *Spaceport Infrastructure Handbook* that was produced under a joint agreement with the national Aeronautics and Space Administration.

William M. Lawbaugh AB; MAT (St. Louis University), PhD (University of Missouri), Postdoctoral Studies at Ball State University, Kent State University,

Mount St. Mary's College. Awards include, First Amendment Award, The Society of Professional Journalists, Sigma Delta Chi, Distinguished Multimedia Adviser of the Year, College Media Advisers, Gold Key Award, Columbia Scholastic Press Association, Columbia University, Eugene Pulliam Award, the Society of Professional Journalists, Pulliam Foundation. His Professional career includes, Assistant Professor, University, Muncie, Indiana; Senior Field Representative, Asian-American Free Labor Institute; Associate Consultant, Technical & Administrative Services Corporation (TADCORPS); Associate Professor, Mount St. Mary's College; Technical Writer & Editor, Smithlain Enterprises; Technical Writer & Editor, George Mason University, Center for Innovative Technology. Recent publications include, Technical Editor, Gateway Engineering Project, Drexel University, *Tools and Tactics of Design* John Wiley & Sons, New York, Technical Editor, *Spaceport Infrastructure Handbook* GMU Publications.

Frank L. Manning BS (University of Michigan), MS (University of Akron) has 20 years of experience in developing, managing and operating expendable launch vehicles. He has recently been involved at NASA in safety and mission assurance having previously served on the aerospace safety advisory panel.

Glen McDougall MSc (London School of Economics), Professional Engineer is an expert in aviation commercialization. He was Deputy Negotiator, ANS Commercialization in Transport Canada where he managed the transition to the private sector from policy concept to implementation. He also specializes in results-based management, performance information and government reform, having held positions as Director General, Business Solutions in the RCMP and as Special Advisor to the Treasury Board Secretariat on a government-wide performance-based reform program. He is on the Board of Directors of the Performance and Planning Exchange, a private-public thinktank on performance management, and former Vice-Chair of the Government of Canada's Management Board on implementation of SAP.

Harold Miller developed the Mercury, Gemini and Apollo Flight Control Simulation and Training System, ran Flight Controller Simulations and Training as a Branch Chief, developed a manpower utilization system for the flight control division of NASA, was technical monitor for all flight controller contracts for 2 years, led the development of space station payload and operational requirements, planning system for space station systems engineering and integration, was involved in the electric drawing system for space station configuration management, and worked in the Shuttle Office of NASA.

John Mulcahy has over 40 years experience in the design, development, test and certification of major propulsion systems, including the H-1, F-1, J-2 and LM

Descent Engine for the Apollo Program and the Space Shuttle Main Engine. Prior to retirement he was at the Office of Space Flight at NASA Headquarters as the manager of Technology Engineering Operations. He began his career as an engineer with the Thiokol Corporation following service with the US Army. Later he worked on the Apollo Project at MSFC and at NASA HDQ, as well as the Delta Project at GSFC and NASA HDQ. In the 1980s he had the lead responsibility in the HDQ Space Transportation Systems office for the Space Shuttle Main Engine technical requirements.

John T. Sheahan BA (Georgetown University), MBA (Columbia University) and a graduate of the National Defense University's Industrial College of the Armed Forces as well as the Federal Executive Institute and the NASA Management Education Program. He joined the staff of George Mason University in 1987 following a 32-year career at NASA Headquarters. At the time of his retirement from NASA, he was the Deputy Director of the Center Operations Division in the Office of Space Flight. He began his career at NASA as a management intern followed by being a program analyst in the Apollo Program Control Office. In 1968 he went to Vietnam on military service. He returned to NASA as a program analyst in the Skylab and Space Shuttle program offices. He later became chief of Program Budget and Control for the Office of Space Flight. In 1982, Mr. Sheahan was selected to serve on NASA's Space Station Task Force where he became the Director of the Business Management Division of the Space Station Program Office. In the late 1980s he was Acting Director of the Space Shuttle Program Office at NASA Headquarters. Among his NASA awards are the Exceptional Service Medal and several special achievement awards. He also received group achievement awards as a team member of the Space Shuttle Return to Flight, Program/Project Management Summer Study, and the Apollo, Skylab, Space Shuttle and Space Station programs.

Roger Stough is the Director of the Center for Regional Analysis and the Transportation and Aviation Policy Programs and Associate Director of the Institute of Public Policy. He holds the Northern Virginia Endowed Chair in Public Policy. He holds a BS in International Trade and Economics from Ohio State University; an MA in Economic Geography from the University of South Carolina; and a PhD in Geography and Environmental Engineering from Johns Hopkins University. Dr. Stough has an extensive publication record, including nearly 200 scholarly and professional publications, books and journal articles. Dr. Stough also has a wide range of graduate and undergraduate teaching experience, with over 25 years experience in eight separate institutions. He has extensive experience interfacing the University with the local, state, regional and national community. He chairs a number of task forces, and advises a wide range of program and university-level policy boards.

Francis T. Hoban

Frank Hoban has spent more than forty years in the aerospace industry. He was born in Minersville Pennsylvania. He joined the US Army serving as a paratrooper with the 82nd and 11th Airborne Divisions. Frank went to school at Parks College of Aeronautics near St. Louis and graduated in 1960. He met his wife Mary Louise at nearby Fontbonne College. Parks is now part of St Louis University. He worked at an airport near William and Mary College in Virginia where he got an MBA. He then joined the Peninsula Airport Commission as its Assistant Executive Director. In 1963 he left airport management for the challenge of National Aeronautics and Space Administration (NASA) at its headquarters in Washington. This was an exciting time to be involved in the aerospace sector and Frank rapidly rose to meet the challenges of his new environment.

In 1970 he was selected to serve on the staff of the NASA Administrator as an Assistant Executive Secretary, first to Dr. Wernher von Braun and later George Low, the NASA Deputy Administrator. In 1973 he joined the newly formed Low Cost Systems Office as the Director of Business Practices. He earned a NASA

Exceptional Service Medal for his work in the low-cost field. In 1976 he was selected as a Presidential Exchange Executive and spent the next year with the General Motors Corporation in Detroit, Michigan.

Returning to NASA, Frank was assigned to an agency-wide task force to implement the historic Civil Service Reform Act. NASA won a major award for its work. As the activities of the task force were winding down, Frank was asked to join the President's Commission (the Kemeny Commission) on the Accident at Three Mile Island as its Director of Administration. The White House, the Office of Management and Budget, the *Washington Post* and the *New York Times* praised the workings of the Commission, which reported in 1979.

In 1982, Frank joined the Space Station Task Force as its fourth employee. In 1986, after the *Challenger* accident, he was tasked to lead an agency-wide program to train the NASA workforce engaged in managing programs and projects. This effort, known as the Program and Project Management Training and Development Initiatives, was very successful and earned him NASA's Exceptional Achievement Medal. It also resulted in a series of books that Frank edited called *Issues in NASA Program and Project Management*. These were not only widely used within NASA to help better management within the organization but were adopted by similar agencies overseas. He also, in conjunction with William Lawbaugh produced two NASA special publications, *Readings in Program Control* and *Readings in Systems Engineering*.

These various publications epitomized Frank's approach to his career at NASA – he was interested in ideas, but equally importantly he was interested in their application. Edward Hoffman of NASA expressed it well when he said, 'Frank came from a project world. He much preferred addressing real issues by working with the best people and staying focused on the customer requirements. When I first started with Frank, he informed me that the more time I spent in my Headquarters office the less effective I would be'.

Frank retired from NASA in 1994. But he did not spend his time idling. In conjunction with Edward Hoffman and William Lawbaugh he set about putting on paper his views on how NASA could improve its internal management. The subsequent volume, *Where do you go after you've Been to the Moon? Case Study of NASA's Pioneering Effort at Change* was published in 1997. It was the first book to analyze NASA's post-Appolo attempt at change, and may actually have been the first analysis of any large government agency's attempt at change through cost control. One reviewer (Mark William in *Earth Space Reviews* said, 'Hoban has succeeded in putting together a well-written and interesting analysis of NASA bureaucracy and culture of the 1970s and 1980s. The book may be history, but its main message – that we can learn from history, especially its mistakes and failures – is as current as ever'.

In the late 1990s Frank began a new career as an adjunct professor and Director of the Continuing Career Program at George Mason University in Fairfax Virginia. This pilot career program began in September 1998 with financial support from NASA, whose employees were struggling with major downsizing. The program teams retired or displaced NASA employees with faculty, graduate students and other university staff to work on aerospace problems and initiatives of concern to government and industry. The university selects the senior research fellows, who receive an annual stipend to supplement retirement or buyout packages. They can spend up to three years in the program, which offers course work as well as consulting opportunities. Fellows in the program come from a wide variety of NASA backgrounds – e.g., Washington headquarters, Kennedy Space Center in Florida and Goddard Space Flight Center in Greenbelt.

The program allowed NASA executives to share their expertise more widely whilst still being engaged in the aerospace community. The group also pursued themes of common interest, and one important output was a report published in 2002 that looked at *Spaceport Infrastructure Handbook*. This pulled together their collective views on the way space transport would develop over the next fifty years or more in a climate where commercial success was as important as national status, defense and scientific research.

Frank put his full energies into the program and was a major success. As Keith Cowing, editor of *NASA Watch* put it, 'He was one of the most engaging, witty, and knowledgeable folks NASA's family ever produced. Frank would often tell me some of the most amazing stories – ones which stretch all the way back to NASA's earliest days. In recent years he was almost obsessed with capturing NASA talent as people walked out of the door at NASA (via buyouts or retirement) and bringing this experience back to NASA on an as-needed basis'.

A similar, but larger sentiment is also expressed in the words of James L. Jennings, Associate Deputy Administrator for Institutions and Asset Management at NASA,

> I have known Frank for about 20 years during my career at Kennedy Space Center. I first became acquainted with Frank through his work with the International Space Station program many years ago. Frank's commitment towards preparing the future generation of NASA's talent cadre was demonstrated throughout his career.
>
> Frank and I both shared a strong commitment to education, so it was of no surprise that our paths crossed over the years as we participated in numerous educational and developmental programs and panels. His work at George Mason University continues to provide opportunities for retiring NASA employees to contribute to the Nation's space program. Frank had a genuine interest in people, and he will be greatly missed across the Agency.

Outside of his professional life, Frank served on the board of Mother Seton School and on a community advisory group at Mount St Mary's College. He was an active contributor to various civic, social and church groups in Emmitsburg and Taneytown. He was also active in a local Catholic high school.

Frank Hoban died on December 5, 2002 at the age of 67.

Chapter 1

Introduction

Kenneth Button

Introduction

Space is often seen by the entertainment industry as the 'Final Frontier'. It certainly poses stimulating and important challenges. Exploring space involves using cutting edge technology and complex management structures. But there is perhaps a little more to it. Just as frontiers reach a point where they have to be exploited, developed and settled, so space has dimensions that extend beyond that of research and exploration in its purest form.

The aim of this collection of chapters gathered as a tribute to the work of Frank Hoban is to look at some of the issues and decisions that have affected aerospace activities over the past decade or so. They include issues of aviation as well as what is sometimes strictly seen as aerospace because the line is a blurred one. There are also many common threads in the types of issues that have needed to be addressed.

The chapters are all original contributions and are authored by those who worked with Frank Hoban. Indeed, some of them provide material that he was directly involved in.

The theme of the collection is not overtly technical but looks at some of the major decisions that have been made regarding aerospace policy. Some clearly have a technical element, but equally there are important economic and political factors that came into play. They reflect an important change in the way space transportation was viewed and whilst the term 'watershed' may be a little strong, the post Moon-landing period did see a shift towards a more commercial attitude to the issue. This is reflected in the papers. Many of them deal with the particular problems the National Aeronautics and Space Administration (NASA) had in integrating private sector activities into its own while at the same time transforming its structures.

There were also major technical developments that needed to be addressed in the 1990s as older technologies and worn hardware needed replacement. This overlapped with the new missions of institutions such as NASA but also with the new demands for services placed on other institutions such as the Federal Aviation Administration (FAA). Changes in the air transportation industry and a rapidly increasing demand for air travel were putting increasing pressures on the navigation and air management infrastructure of the US.

The 21st century has seen new challenges emerge but some of the older ones still remain. These can only be touched upon in passing. The end of the Cold War saw a reduction in the military dimension of space but with the advent of potential new threats, this dimension of aerospace activities is reemerging.

Perhaps more germane to this volume, there are new questions being asked, especially after the Columbia accident, about the importance of manned as opposed to non-manned space travel. Tied with this are issues about the roles of the public and private sectors in aerospace activities. Research had traditionally been largely kept isolated from production. The public role of institutions such as NASA changed in the 1990s. NASA for example began to engage in production as well as pure research, but this seems now to be in question.

Linked to this are issues of internal management and management cultures of institutions such as the FAA and NASA. The August 2003 report on the loss of the space shuttle Columbia highlighted some of the managerial problems within NASA as well as the specifics of the particular incident (Columbia Accident Investigation Board, 2003). Issues of culture and management style are given as much space as those of technical matters of tile materials and escape possibilities. In its press release the point is made succinctly:

> The Board determined that physical and organizational causes played an equal role in the Columbia accident – that the NASA organizational culture had as much to do with the accident as the foam that struck the Orbiter on ascent.

The remainder of this Introduction is intended to offer both a general background to the challenging world in which the case studies fit and some context to the particular contributions.

Background to Developments in Aerospace

The space program in the US grew from visionaries such as Robert Goddard who launched the first liquid-fueled rocket in 1926 and who went on to incorporate many of the features of modern rocketry into his equipment. The Second World War saw the introduction of the German V-2 rocket as a weapon of mass destruction. The rocket weighed 14 tons when fueled and generated 25 tons of thrust.

The post-war era saw the continued development of rockets as delivery systems for munitions but it also witnessed their gradual emergence as scientific instruments and as commercial entities. The US program really began in 1946 under a team lead by Werner von Braun and was militarily oriented, making use of captured and modified German hardware. Whilst there was some effort by the army to get a satellite into orbit, the first orbiting satellite was Russia's Sputnik launched in October 1957. The US launched Explorer 1 three months later using a Redstone Rocket.

The emergence and gradual intensification of the Cold War saw space become a political battleground with both the US and the USSR seeking to show their prowess as super-powers by developing commercial as well as military space programs. In October 1956 to this end the US established NASA as a US agency dedicated to the peaceful exploration of space. Three years later, the US put its first astronaut into space for a sub-orbital flight aboard a Mercury Redstone Rocket. This was the stimulant, combined with earlier USSR orbital flights, to begin the program to safely land astronauts on the Moon and to bring them back safely. The aim of the program was to give the US pre-eminence in space. The practical result was the Apollo program.

The Apollo program was as much an exercise in logistics as it was in engineering. Certainly new rockets were needed to propel astronauts beyond Earth's orbit, to land on the Moon and to return. But additionally a spaceport would be required for assembly of the rocket and for the launch. To this end, 90,000 acres of land in eastern Florida was developed adjacent to Cape Canaveral Air Force Station. In December 1963 the NASA portion of the complex was renamed the John F. Kennedy Space Center.

The facilities required to handle the rocketry of the Apollo program were massive with the Saturn V rockets being 36 stories high and generating 7.5 million pounds of thrust to get its 150-ton payload into low-Earth orbit. Extensive barge facilities were needed to get parts of the rocket to the launch site. The $800 million, four-year civil engineering construction program was in itself a major feat of civil engineering.

The post moon-landing role of NASA, and the space program more generally, has been somewhat different. The end of the Cold War led to cooperation and the development of the International Space Station. The US approach to this, and other aspects of space transportation has been the parallel development of reusable manned space launchers to build and serve the space station, combined with unmanned, robotics technologies for incursions into deeper space. The Space Shuttle program, of which the first flight was in 1981, was an integral part of the former, although it was not the only option considered (Comptroller General, 1973).

The US began to move towards a more commercially driven approach to space in the 1970s. Initially this involved NASA up-grading its Delta launch vehicle (see Chapter 3 in this volume) for use by RCA's communications satellite program, SATCOM. To day a number of nations have the capacity to launch satellites and it has become a major business.

US policy makers have encouraged this with enabling legislation over the years. In particular the Commercial Space Launch Act of 1998 was passed with the explicit purpose of developing a commercial space industry in the US. It was aimed at improving opportunities for commercial launch providers. It states:

> Except as otherwise provided in this section, the Federal Government shall acquire space transportation services from United States commercial providers whenever such services are required in the course of its activities. To the maximum extent practicable, the Federal Government shall plan

missions to accommodate the space transportation services capabilities of the United States commercial providers.

Commercialization has also come in another form the ever-increasing involvement of the private sector in the development and construction of spacecraft for the US government. One aim of the policy is to reduce the financial burden of space transportation borne by the taxpayer. To this end for example, following the 1995 *Kraft Report* (Space Shuttle Management Independent Review Team, 1995) the United Space Alliance, a joint venture between Boeing and Lockheed Martin, has become the prime contractor for the Space Shuttle. Some 10,000 employees are involved and perform a variety of functions from astronaut training to flight operations – many of them used to work at NASA. NASA also contracts for specific tasks – Lockheed Martin makes the external tanks and applies the thermal protection to them. The policy has not been entirely successful however, and a planned saving of $1 billion a year from 1995 has turned into an actual savings of $1 billion over the first six years of operations.

It is this issue of the economics and the respective roles of the public and private sectors in space transportation that we now turn to. It is at the heart of many of the decisions and policies that are discussed in this volume.

The Changing Economics of Space

The space industry is big business and it is increasingly an international business. In 1999 it was estimated by the Federal Aviation Administration that US economic activity linked to the commercial space industry totaled over $60 billion, of which commercial space transportation was either directly or indirectly responsible for over $16 billion. It is a major employer with nearly 500,000 jobs either directly or indirectly resulting from it. It is not surprising, therefore, that there is often competition between locations to be involved in the sector (see Chapter 8 in this volume).

Jobs in the aerospace sector are neither typical in the remuneration that they bring employees nor in geographical distribution. The US aerospace and aviation industry employs about 2 million workers directly with wages that were in 2000 35% higher than the US average – although this was down from the 43% differential of 1996. The best paid work in the guided missiles and space vehicles industry, satellite communications, and search and navigation.

While the industry generates high incomes, there are major regional differences in the US as to where employment is located and the local wage levels (Commission on the Future of the United States Aerospace Industry, 2002). Controlling for population size, for example, the states with the higher levels of aerospace and aviation employment in 2000 were Washington, with 44 workers in the sector out of every 1,000 jobs, and Kansas with 43 per 1,000. But the highest nominal wages were to be found in the District of Columbia with Connecticut and Washington following. In terms of more specific activities, Texas, the home of the Johnson Space Center, led in space research employment, followed by Maryland

and Virginia. Alabama came fourth with some 2,600 space research employees at the Marshall Space Flight Center.

In the days of the Cold War there was an inevitable need for the US to focus its space program on defense and on prestige projects. Whilst there is still a need to recognize these factors, the globalization of trade and national economic needs has brought to the fore the role of the aerospace sector as a driver for economic prosperity. It has also become appreciated that providing for the more traditional roles of the sector can be more effectively achieved through new structures and institutional designs. In particular, there is a need to rethink and rebalance the roles of the public and private sectors.

Extreme libertarians or centralists often see the arguments for public and private involvement as black and white. The former group argue for minimalist government involvement while the latter are wary of market forces. In practice, there are many refinements to consider and in practical terms the divide between the public and private sectors becomes one of degree. The classical economic arguments put forward by Adam Smith are that the market should normally be allowed to prevail except in some very prescribed circumstances. In the case of space transportation, the most frequently argued case for public supply, other than for military reasons, is when research and development, and delivery, are too expensive for the private sector to invest adequately. Only governments have the resources to do an adequate job.

But there are counters to this and these were used extensively in the 1970s in aerospace debates. First, space transportation is not a unified activity – it can be unbundled into its various components (see Chapter 6 in this volume). Unbundling allows for smaller individual investments by the private sector in the form of sub-contracting and outsourcing. Consequently, while the government or its agent (e.g., NASA or the FAA) lays down the overall strategy it buys in those parts of the activity that can be provided most efficiently (and usually more cheaply) by the private sector. The aim of the approach is one of simple cost effectiveness, it does little to stimulate new initiative or draw upon the entrepreneurial spirit of the private sector. There are also practical problems of management and the awarding of contracts, especially in situations where there are a limited number of potential private sector suppliers. Matters of monopoly powers arise. There are also situations where it is difficult to specify the exact output required from the private sector supplier, or where specifications change over time. Contracts lack specificity in such cases, and the private sector supplier thus has less incentive to minimize costs.

A more integrated approach is where the agency gives over a large part of its activities to the private sector. This is effectively what has happened in the context of the Space Shuttle. The private company takes a much larger role in the overall design and execution process. Economists often favor such an approach in the case of a fairly mature product where the production and execution has become virtually routine. It is less likely to meet its objectives of both quality control and economy when the product is in the development stage or is subject to continual up-dating. The privatization of the US Space Shuttle program from 1995 followed decisions based on the *Kraft Report* that no further significant refinement should be made to the Space Shuttle. It was premised on the argument that the shuttle was 'a mature and reliable system...about as safe as today's technology will provide'. This is now being questioned.

More genuinely private sector is when purely commercial companies develop space vehicles. The infrastructure may be public, although this need not be the case (see Chapter 7 in this volume), but the vehicle may be the product of private enterprise. For example, commercial air transportation grew as the result of private sector initiatives with suitable rewards for 'winners'. The Wright Brothers flew, took out a set of patents and then developed their technology further. This continued until the US rescinded their patents in the First World War for defense reasons. Prizes have also been successful in stimulating investment that led to technical developments that at the time were revolutionary – Charles Lindbergh flew the Atlantic in a quest to win the Orteig Prize. Once the private sector has invested then the public sector (as well as the private) may lease or buy the product (see Chapter 4 in this volume).

Whatever the pros and cons of privatization and its various forms, the change in focus began in the US in the mid-1970s and has grown subsequently. The change has been dramatic and by 2002 about 85% of NASA budget, about $13.3 billion, actually went to private contractors. The real economic issue is now to ensure that these funds are spent efficiently and that an optimal flow of private sector money is attracted into aerospace activities.

The Chapters in this Volume

The chapters in this volume reflect the great diversity of factors that can affect decision-making. They are not comprehensive in that there are certainly many additional case studies that could offer further insights, but their coverage is still significant. They are also different to many other papers that cover similar themes in that in many instances the individuals who wrote the case studies had themselves a direct involvement in the program or policy under review. Some readers may indeed call several of the contributions 'war-stories'. But personal reflections and personal experiences are important in improving our understanding of how decisions are made and are ultimately carried through. Decision processes, however much management scientists like to treat then analytically, almost inevitably contain a personal element.

There is often a tendency in this type of book to highlight successes. This one certainly does contain success stories, but it also contains chapters on failures and on unresolved issues. Learning is not only a process of seeing how aims can be attained but also it is about how future mistakes can be avoided, or at least their adverse impacts mitigated.

In terms of presentational style, the papers are also deliberately designed to be accessible to a wide audience – they are not written only for those with knowledge of science but rather for those who have an interest in the management of decision-making. They are in prose and not equations. It is also hoped that readers will find the topics covered relevant to many of the on-going debates about aerospace policy in the 21st century.

Some of the chapters deal directly with aviation and aerospace infrastructure. The matter of how air navigation and air traffic-management services are provided is not only important but also topical. The material here covers both ownership issues (Chapter 2) and technology change (Chapter 9). The former has taken on a particular relevance since the corporatisation of the Canadian air traffic system with other countries such as the UK also moving to a more commercial approach for the supply of such services. The case of technical change, and the types of problems encountered by the FAA when trying to initiate the microwave landing system has relevance to the European Union which is seeking to develop a Pan-European Single Skies Policy within its borders.

But any form of transportation infrastructure also embraces hubs – traditionally rail terminals, airports and the like. In the aerospace context the equivalent are spaceports (Chapter 7). These have in the past involved significant public investment but are now taking a diversity of forms. There are inevitable decisions to be made about how many are needed, what forms they should take, and the extent to which they should be financed with public money. The old policy debate of the mid-19th century about who should own and operate the rail track and who should own and operate engines and rail cars has its 21st century counterpart. The debates about the best way to run railways continue, and no doubt those regarding spaceports will continue to run. The hope here is that at least the dimensions of the decisions that need to be made are clearly articulated.

Launch vehicles tend to attract the majority of scientific interest and certainly arouse the largest public emotion. The collection contains a number of contributions that both cover general issues of seeking-out the most efficient type of launch vehicle, defined in this case as the least cost way of meeting pre-defined objectives (Chapter 6). But there are also individual programs to consider.

Some of these initiatives, such as the Delta program that, besides other things moved the US into the commercial aerospace age, were highly successful (Chapter 3). Others, however, were less successful but provide useful guidance as to the problems of developing space technology and activities. The X-33 program (Chapter 8) was a pioneering public/private initiative intended to provide a vehicles (VentureStar) to replace the Space Shuttle amongst other things. Leasecraft (Chapter 4) was a private sector initiative aimed at offering commercial satellite services but failed to meet commercial criteria.

Finally, the image of air transportation and space travel conveyed in films and pictures is one of technological cleanliness. The reality is far from this. Enormous amounts of energy are used and various pollutants are released into the atmosphere. As the chapter on spaceports reveals, air and space travel require large terrestrial land for their landing and take-off sites. These factors, externalities to economists, tend to be ignored in much of the technical debate. The coverage in this volume is not comprehensive but Chapter 5 does provide a discussion on policy developments regarding noise suppression from aircraft. This is not an insignificant social problem.

References

Columbia Accident Investigation Board (2003) *Report*, CAIB, Washington, DC.

Commission on the Future of the United States Aerospace Industry (2002) *US Aerospace and Aviation Industry: A State-by-State Analysis*, US Government Printing Office, Washington DC.

Comptroller General (1973) *Analysis of Cost Estimates For The Space Shuttle and Two Alternate Programs*, Report to Congress, Washington DC.

Space Shuttle Management Independent Review Team (1995) *Report*, NASA, Washington DC.

Chapter 2

The Privatisation of the Canadian Air Navigation System

Glen McDougall

Introduction

In 1996 The Government of Canada sold its air traffic control System to NAV CANADA, a private, not-for-profit company created especially to own and operate the system. This was the first time in history that any country had privatised its national air traffic control system. To date, Canada remains the only country in the world with a completely privatised air traffic control system, called the Air Navigation System (ANS).

Canada's achievement, which involved a sale price of C$1.8 billion and the transfer of over 6000 government employees to the private sector, would not have been possible without the development of a unique model for privatisation. This model evolved from lengthy public and inter-departmental consultations that were essential. In reconciling conflicting stakeholder interests and resolving numerous complications. Particularly challenging were:

- There was no international precedent for such a move that could provide guidance.
- The ANS was not a separate entity but was integrated within its department (Transport) and other government departments (e.g. Justice, Public Works).
- The ANS was supported by several hundred million dollars per year of government appropriations, making the entity appear to be commercially non-viable, and unattractive to potential buyers.
- Financing a private start-up corporation without equity had not been done before on this scale in Canada.
- Most employees were unionised and were represented by nine different unions, making it difficult to find transfer terms and conditions that would satisfy everyone while giving the new company maximum commercial freedom.
- Many groups, including aircraft owners and pilots, as well as some key senior government officials, opposed moving an essential government service into private hands.

This case study describes the policy development and consultations that led to the privatisation decision, the implementation process and the current status of air traffic control as provided by the new corporation.

The Policy Decision

The Pressure for Change

The Canadian government changed its air policy in the mid-1980s to deregulate airline activity, with the consequence that there was tremendous growth in air traffic. This was particularly true at locations such as Toronto where hubbing was taking place. At the same time there had been an effort on the part of government to downsize operations and reduce budgets in many areas, including the air navigation system.

This combination of growth in workload and reduced financial budgets created a shortage of air traffic controllers at key locations. The government human resource system did not provide the flexibility to relocate staff, reallocate workload, or give adequate incentives to make working in high pressure, demanding locations attractive to employees. In addition, training of new air traffic controllers was halted as a budget saving measure. The result was a reduction in service to air carriers, major delays, and restrictions placed on the capacity at an airport (hourly flow of aircraft) because of air traffic control constraints. The restrictions in Toronto, in particularly, constrained departures and arrivals at airports that connected to Toronto. As a consequence, the late 1980s saw major delays and costs imposed on the air transportation sector, and significant inconvenience to travellers.

The controllers' union claimed that safety was being affected. The shortage of controllers required large amounts of overtime be worked to keep the system operating. There was concern that excessive overtime was leading to fatigue on the job. Efforts by managers to remedy the situation by reducing services in less critical locations and relocating staff were often resisted by communities and their elected representatives.

Something had to be done (Poole and Butler, 1996). One solution to this dilemma was to extract the air navigation system from the government process of budget cuts and imposed services, remove government constraints on such matters as personnel policies, eliminate the dependence on tax revenue and establish independent safety oversight. In other words, to reorganise the ANS to be more independent of government and to commercialise that entity to be financially self-sufficient.

Several stakeholders reached this conclusion simultaneously. Senior management within ANS had proposed this solution. Air carriers made it known to the government that they were unhappy with the current arrangements for ANS and recommended that commercialisation options be studied. The air traffic controllers' union also recommended commercialisation, in part to gain freedom from perceived government restrictions on their bargaining rights.

The Canadian air carriers obtained the support of the commercial pilot association, the business aircraft association and the air traffic controllers' union. Together, the four groups recommended that the Minister of Transport study commercialisation. A Minister's Task Force on Aviation Matters and a Royal Commission on Passenger Transportation also concluded that there was a pressing need to study commercialisation of air traffic control.

Building Momentum

Despite pressure from within and outside to look at commercialisation, there was resistance to the change being proposed. Senior Transport officials remained cautious and resistant. In their view the public expected ANS to be a government service and, in addition, employees could be adversely affected. The central agencies, however, supported the concept and conducted analysis that confirmed their position (Transport Canada, 1994a; b; c; d). The final deciding factor was the Minister of Transport, Doug Young, who, after briefings by departmental staff and the aviation industry in January 1994 on the options, decided to explore commercialisation. This was endorsed by Cabinet and was announced in the government's annual Budget speech a few weeks later.

The private pilot association also hesitated to endorse the idea. It feared that general aviation would have to pay exorbitant fees each time a private pilot contacted a tower or used an ANS service.

There were other stakeholders' concerns: Would a commercialised ANS mean higher costs? Would services to the North and remote communities be decreased? Would bilingual requirements continue to be met? Would the military suffer a loss of sovereignty over domestic airspace with a resultant increased security risk? Would costs to the travelling public increase significantly?

Once the government decision had been made to explore commercialisation, Transport Canada began the process by setting up a small team (A director – the author – and two officers) under the Assistant Deputy Minister, Aviation to prepare necessary information and to analyse the benefits and concerns of commercialising the air navigation system. The team, joined by the Director General of Air Navigation Services, met with the heads of a number of domestic organisations and international air traffic control units that were already commercialised, to examine the experiences, benefits and problems with the new organisations. They also reviewed existing literature, and conducted special studies with the cooperation of internal financial and economic analysts in the department, and with external aviation consultants. An analysis of insurance requirements and costs was completed with a major London brokerage house specialising in aviation insurance.

The team exchanged information frequently with colleagues in the US Department of Transportation and the FAA, and consulted with the governments of New Zealand, Australia and Germany on a regular basis. The Canadian Department of Finance provided ongoing advice on the financial matters.

The team examined organisational options, international experience, revenue sources, service and cost impact on users, impact on employees, potential user fees, financial impact on the department, the need for safety and economic regulation, capital investment scenarios and implementation suggestions.

The team obtained sufficient evidence to convince the government that public consultations were warranted. Thus prepared, the government announced its decision in early 1994 to consult the public on the potential to commercialise Transport Canada's major operations, including the Air Navigation System.

Public Consultations

The transportation department set up an extensive consultation structure. It hoped to arrive at a consensus by the end of the year on whether or not to commercialise ANS, and if so, which organisational form would be best. An advisory committee was established headed by the Assistant Deputy Minister, Aviation. This committee included external clients and stakeholders of ANS plus the nine unions that represented ANS employees, as well as the associations representing national air carriers, regional air carriers, commercialised airports, aerospace manufacturers, commercial pilots, business aircraft operators, private aircraft owners and operators, and the Transportation Safety Board.

A steering committee was also established, chaired by the same Assistant Deputy Minister, with membership from senior management in the department including operations, policy, personnel, and finance. There were in addition representatives of the Privy Council Office, the Treasury Board Secretariat, the Department of Finance, the Department of National Defence and the Department of Environment (weather service).

The advisory group established study groups of operational and functional specialists, as well as external stakeholders on themes of finance, personnel, implementation, legislation, public affairs and policy. These groups completed studies and analyses of specialised issues and reported back to the advisory committee.

The government produced a series of consultation documents in cooperation with the advisory and steering committees. These documents made it clear that there were two overriding principles: the government would not allow commercialisation of the ANS to have a negative impact on the safety of the travelling public, and that commercialisation could be effected in a number of different ways, from special agency status to privatisation. The objectives for commercialisation included:

- The ANS should be able to respond better and more quickly to user needs.
- The ANS should be able to achieve more rapid improvements in internal and transportation efficiency.
- Dependency on taxpayers should be eliminated as soon as possible.
- The ANS should operate in a business-like way.

The institutional framework for commercialisation would have:

- the freedom to manage resources and people effectively;
- the ability to make efficient procurement decisions based on commercial principles;
- access to the capital markets to borrow or deposit funds;
- the ability to rapidly adopt new technologies;
- funded by those making use of its services;
- a business-like approach and discipline in operations and decision making;
- accountability to owners and customers.

Public consultations were held across the country. Different organisations such as regional economic councils, municipal authorities, colleges and aviation councils hosted these 'town hall' sessions. They were responsible for transportation arrangements for the facility, publicity, and media attention and so on. The department presented the proposal and discussed options with the audience. Copies of the proposal and several theme papers were distributed before the meetings.

Employees were also surveyed to solicit their views and concerns.

This extensive consultation process formed the basis for the model's success. The Department received compliments on the wide and thorough consultation process and, most importantly, no stakeholders came forward after the consultations to indicate that they had been left out.

Background Studies

The study team defined five key areas to be addressed before consulting the public. The reports covered:

- Principles and Options for Commercialisation: including discussion of the role of government and the public interest in air navigation services;
- Safety Regulation: separating the regulator from the provider of services;
- Economic Regulation: protecting users from the monopoly created;
- International Experience: a summary of commercialisation in other countries prepared by the Airways Corporation of New Zealand;
- Illustrative User Charges: examples of what a new commercial entity might charge for air navigation services.

The study team hired consultants as necessary to complete this work quickly. The reports were widely distributed to facilitate the debate of the options and findings during the consultations.

Additional studies included: market risk assessment (i.e. the impact of future changes in the aviation industry on revenues); the impact of commercialisation on capital funding; historical and forecast financial statements; boundaries of ATC (carving the organisation out of the public service), potential for internal efficiencies, the estimated financial value of the new organisation, insurance costs and liability mitigation, corporate governance and accountability, principles for legal formation of a new corporation, and fundamental principles for the creation of an air traffic control corporation.

The Public Interest and the Role of Government

A deciding factor on the choice of organisation would be the ability to meet the public interest. Transport Canada officials worked closely with the central agencies of government to help define the key elements of the role of government and the public interest, discussed these at the steering and advisory committees, and included the results as a section in the public consultation document.

Transport Canada officials explored the concept of 'steering rather than rowing.' This is the principle where government sets policy and regulations, and empowers others to carry out the operations. Following this principle, the public interest would define what services should be provided and how the public should be protected, leading to determination of the role of government rather than starting from the premise that the government role was pre-defined by declaring that ANS was an essential government service.

The consultation documents identified three key elements of the public interest. These were: requiring that the ANS organisation exist (i.e. it could not be allowed to go bankrupt or cease business), that it be safe, and that it contribute to national transportation efficiency. The public interest also required that the air traffic control system:

- provide equitable access to all users;
- maintain a national airspace structure that meets government and public needs;
- provide appropriate services to remote communities;
- ensure that international obligations are fulfilled;
- contribute to national sovereignty and security;
- not abuse its monopoly position;
- remain nationally owned and controlled.

Each item of the public interest was explained in detail in the consultation documents. For example, the requirement on national airspace structure addressed the government's obligations under the Chicago Convention of 1944 and the sovereignty of states over the airspace above their territories. This included provisions for bilingual airspace over parts of Canada.

During consultations, there was no debate on the description of the public interest as presented by Transport Canada, and the definition set out above was accepted.

All stakeholders agreed that many organisational forms could meet the public interest requirements. The government did not need to own and operate the ATC system to achieve its public policy goals. Safety and security were not guaranteed by keeping government as operator of the ANS, as these conditions already depended on many private sector organisations and individuals including pilots, ground crew, manufacturers and airline operators. The federal government could use its powers of legislation, regulation, enforcement, appeal, delegation, subsidy and so on to require or induce a commercial ATC organisation, of any form, to respect the public interest requirements.

Most importantly, all agreed – including Transport Canada's regulators, the Transportation Safety Board, the air carriers, the air traffic controllers union and the commercial pilots association – that safety of air travel would not be compromised by commercialisation of the ANS. The government would regulate and monitor safety of a commercial air navigation system as it did with aircraft manufacturers and air carriers. In fact, safety officials generally believed that safety might be improved by the separation of regulator and service provider.

Organisational Options

In designing a new organisation to manage the air traffic control system in a commercialised setting, several options were put forward for public consultation. The four government-owned options included: the creation of a separate, commercial organisation within the Department of Transport (a special operating agency), contracting for the operation of the service (government-owned, company-operated), and two types of state-owned enterprise – appropriations-dependent and financially independent.

The three privatised options consisted of a mixed enterprise where the government held some shares in a private company, a non-share capital, not-for-profit, private corporation and a share-capital, for-profit, private company.

The government had no preference as to the form of organisation chosen, provided the new organisation met the public interest. Instead, it relied on public consultations and user preferences to guide the final outcome. Consultation documents stated that choosing an option was a matter of both pragmatic concerns and finding a consensus among stakeholders and the public on which approach was likely to best serve their interests. Again, inclusion was key to the success of the process.

One factor influencing the final choice was that a commercial entity would collect revenues from user fees for the use and availability of the air navigation system rather than depend on finance from taxes and general appropriations from government.

The air carriers stated that when users paid for these services, they should have more say in the delivery of those services (user pay, user say). This implied an accountability shift from taxpayers and their elected representatives to the users – the air carriers (who would pay most of the costs), business aircraft and general aviation. The governance structure of the new organisation needed to reflect this reorientation of accountability.

Of the seven options considered, the Advisory Committee selected three for further study and public consultations: a financially independent government corporation (Crown Corporation in Canada), a mixed enterprise, with joint ownership by the government and private interests, and a not-for-profit (non-share capital) private corporation.

The consultation documents included detailed description of each organisational option. Some of the concerns that surfaced during the consultation phase on organisational choice were:

- The airline industry preferred only privatised options be considered to ensure that the new organisation was independent from government budget decisions, direction on services and personnel policies.
- Industry was concerned that if a government corporation were established as an interim step, it could be many years before privatisation occurred.
- Most of the nine unions preferred the status quo, or if commercialisation were to proceed, a government corporation; they felt that government should move one step at a time, and that their members would be better protected if the company were still government-owned.
- The air traffic controller's union wanted the new organisation to be completely independent of government, and preferred the not-for-profit, private corporation. This union was very interested in privatisation, partly to regain the right to strike, but also to bargain for some wage movement since federal public servants were subject to a six-year wage freeze at the time.
- Central agencies preferred the government corporation option. They believed that there were too many issues to resolve in moving directly to a private corporation, and that the organisation should first be formed and financial statements produced before privatisation could begin. To ensure that the government corporation did not become permanent, they proposed that the legislation creating the government corporation include sunset provisions that specified that privatisation must follow in a pre-established time-frame.
- The military had no preference as to the form of organisation, but were adamant that the support of their Minister in Cabinet would rest on the protection accorded the National Defence interests.
- Some northern and remote communities feared that their services would be reduced if purely commercial interests governed decisions on service levels.
- The international aviation sector was concerned that the new organisation, if independent of government, be required to comply with international agreements. They also preferred that the new organisation be independent of government budget controls and directions on services.

No one expressed interest in the mixed enterprise. Consequently, the choice of organisational option came down to two: the state-owned enterprise and the not-for-profit, private corporation.

Consensus and Decision

Pulling together a coalition involved taking into account many diverse interests. The air carriers recommended that the new organisation be a not-for-profit, private corporation. Although they wanted the new organisation to be privately owned and operated, i.e., independent of government interference, they were concerned that a for-profit company could charge monopoly prices at the airlines' expense. On the other hand, some such as, the Department of Finance, believed that the profit motive could lead to increased efficiencies that could lower the overall cost of services.

The Air Transport Association of Canada, representing the various Canadian air carriers, worked to build a consensus around the not-for-profit option. They made constant efforts, often behind the scenes, to ensure that the associations and unions supported a common approach to the future of the Air Navigation System.

The unions eventually supported the airlines by accepting (not preferring) privatisation as a not-for-profit corporation. The exception was the air traffic controllers' union that advocated that option from the beginning. In accepting the not-for-profit option, the unions demanded only that the rights and privileges of their members be protected under the new organisation and that financial viability be demonstrated.

The Canadian Owners and Pilots Association, representing general aviation, had difficulty with the privatisation option. It joined the consensus for awhile, but later returned to its position that the new entity should remain government controlled. This, it believed, would best protect the interests of members in keeping fees to a minimum, with full access to services, especially since pressure could be brought to bear through its lobbying powers.

The central agencies of government preferred either the government corporation as an interim step (to ensure organisational start-up stability) or a full for-profit option (to maximise efficiencies) through a share sale. However, they accepted that a stakeholder consensus was strongly in favour of establishing a not-for-profit corporation, and supported this recommendation.

The consensus was built. The Advisory Committee recommended to the Minister of Transport that the air navigation system be privatised as a not-for-profit, private corporation. All members of the Committee expressed satisfaction with the consultative process, agreeing that the results were unprecedented in allowing industry, government and labour to contribute constructively to the development of a new organisation for the ultimate benefit of the public.

The Ministry of Transport prepared an outline of 38 principles that were then discussed with the Advisory Committee and when finalised, formed the basis for the recommendation to Cabinet for ANS commercialisation. Cabinet considered the recommendation and in early 1995 announced a decision to privatise the air traffic control system as a non-share capital, not-for-profit corporation.

Implementation

Establishing the New Corporation

The government considered various ways to start up the new corporation, but chose to facilitate the establishment of a special-purpose, not-for-profit corporation that would negotiate with the government for the transfer as had recently been done in divesting airport operations. Several members of the advisory committee were asked to form an interim Board to establish a not-for-profit corporation under existing Corporations legislation with the purpose of purchasing and operating the ANS (government was forbidden by legislation to establish private corporations).

The government worked with this new group to develop the letters of patent and corporate by-laws for the new organisation, and required that the name chosen be bilingual (French-English). NAV CANADA was the name chosen by the interim Board, and the corporation came into existence under the Canada Corporations Act in the spring of 1995.

The government specified the structure of the organisation. Since there would be no shareholders, the Canada Corporations Act provided for members who could nominate Board directors. The members are the national association representing air carriers; the business aircraft association; the NAV CANADA association of unions; and the Minister of Transport. The 15 board directors comprise:

- three directors nominated by the airline association member;
- two directors nominated by the business aircraft association member;
- two directors nominated by the union association member;
- three directors nominated by the government member;
- four directors selected by the ten directors above;
- the CEO of NAV CANADA.

The CEO may not also be the chairperson of the organisation; otherwise the directors are free to elect the chair from among themselves. Other restrictions prevent elected officials, public servants, clients or users from being on the board. The Dey Report commissioned by the Toronto Stock Exchange was used to determine the governance structure (Dey, 1984).

To entrench these and other restrictions, the Minister of Transport approved the letters patent and certain provisions of the corporate by-laws; these provisions can only be changed with approval of the Minister. The new corporation was now empowered to negotiate the purchase of the system from the government and the transfer of employees and assets.

Negotiations

NAV CANADA engaged two law firms to negotiate with the government on its behalf, one specialising in corporate law, the other in labour law. It also engaged financial advisors.

The government established a dedicated negotiating team of public servants with specialisations in finance, administration, operations, personnel and legal with contracted assistance from a commercial law firm, a human resource firm, and Canadian and American financial advisors. The negotiations were headed by a Chief Negotiator – Paul Gauvin, Senior Assistant Deputy Minister (Finance) in Transport Canada – and the team was headed by the Deputy Chief Negotiator – Glen McDougall.

The government also established an Implementation Team comprised primarily of employees who were familiar with the ANS operations, assets and personnel and who would be moving to the new corporation. This team had the technical expertise to prepare material for the due diligence process, where the potential purchaser could examine records on such items as land holdings, lease arrangements, assets, agreements, and contracts. This team reported to the Assistant Deputy Minister, Aviation.

An Agreement in Principle was reached in December, 1995, that established the sale price and identified terms and conditions that needed to be negotiated or satisfied before the sale and transfer could take place, including a condition that the legislation would reflect specific principles as annexed to the Agreement.

The negotiations were intense, complicated and very detailed. Like any commercial acquisition, the buyer had to be certain of every aspect of the purchase and its future responsibilities for operations. However, there were many and diverse risks to be considered. The details of legislation and departmental agreements were not yet developed; operational obligations (public duties) were still to be defined; and major new user fees were imminent but not yet flowing. In addition, there was no real inventory of land holdings or accurate databases of electronic assets and buildings. Even the exact number of personnel to be transferred was not known.

The extensive number of items to negotiate resulted in delays in the transfer from a target of several months to about 18 months (October 31, 1996). Examples of the negotiated items are listed at Annex A. Even by the date of sale, not all items had been resolved. Consequently, the government entered into a large undertaking with NAV CANADA, where remaining tasks were identified, primarily relating to land transfers, airport agreements and environmental baselines and clean-up.

Determining Financial Value

Two firms were hired to provide financial advice to the government. These were Nesbitt-Burns of Toronto and Goldman-Sachs of New York. Price Waterhouse provided accounting services.

All three firms recommended that the sale be based on a value determined on the net present value of the income stream (revenue over costs) instead of on asset value. Total annual revenues for the business, primarily from user fees that were yet to be introduced, were estimated at close to a billion dollars Canadian. At the time of initial valuation, the organisation derived its revenue primarily from the air transportation tax (commonly known as the air ticket tax, a fee based on ticket price paid by passengers when they purchased their tickets), from user fees for transiting Canadian-controlled, international airspace, and from general tax appropriations.

Following estimates of costs and revenue for the new organisation provided by the accounting firm, the financial advisors produced an estimate of the range of value. Before the negotiations commenced, this range had been publicly stated as between C$1.1 and C$1.3 billion.

Initially, it was questionable whether the ANS would attract any price at all, since it was losing money. However, financial projections indicated that air ticket tax revenues would exceed costs in several years. It was also recognised that the government was selling monopoly rights, as it was not practicable to allow competition in the provision of air traffic control. (Some considered that National Defence provided a level of competition by controlling pockets of airspace, either themselves or through contractors; however, it was not envisaged to expand this service to non-military providers.)

During the consultation phase, the government estimated user fees under several scenarios for the new corporation. These were based on traffic flow and weight of aircraft and were consistent with International Civil Aviataion Organization (ICAO) guidelines and international practice. The resulting price estimates compared favourably to prices charged for similar services in other countries that operated at full cost recovery. This provided assurance to NAV CANADA during negotiations that it would be feasible to replace air ticket tax revenues with reasonable user charges.

Also during the consultation process, several participants commented that the government was missing a significant source of revenue, in that air traffic services were provided to flights over-flying Canada, at no charge. The cost of providing service to these flights was significant, as there are a large number of international flights transiting Canadian airspace between Europe/Asia and the United States. These flights must be monitored, separated and often guided on their descents or ascents to/from northern US airports. In late 1996, the government introduced over-flight fees which, at full strength, would generate about $200 million extra per year, putting revenues, for the first time, about equal to costs.

The improvements in the revenue prospects for the new corporation helped improve the government's negotiating position. NAV CANADA opened the bids with a low offer. It stated that there were many uncertainties and risks with the new organisation. It had never operated as a business, there was no track record of revenues and costs, there was a high-risk major technical procurement underway that had experienced delays and cost increases, there was no example internationally of a similar private business and so on. It was also recognised by both sides that although this was a monopoly business, revenue would be constrained. These constraints would come from international agreements that limited revenues to full cost recovery, recognition that an economic regulatory regime would be put in place to limit price increases and that price sensitivity would place practical limits on revenue.

Projections of revenues over costs under various scenarios formed the basis of price negotiations. NAV CANADA and the government settled on a sale price of C$1.8 billion gross, with a cash component of $1.5 billion and the remainder comprised of certain financial obligations of the government that were assumed by NAV CANADA.

The history of offer and counter-offer shows that both parties were reaching the limits of their acceptable ranges, leaving nothing on the table between this willing buyer and willing seller. Both sides considered terminating negotiations when it appeared that price agreement could not be reached, but delicate negotiations involving the Deputy Minister of Transport and the Chair of NAV CANADA eventually produced an agreement.

During negotiations, each side had fall back positions. NAV CANADA, having no equity, had struck an agreement with the Canadian government to be reimbursed for reasonable expenses should negotiations have failed. The government had the option to commercialise the ATC in a form other than a not-for-profit corporation.

Financing a $1.5 Billion, Non-share Capital Enterprise

The financial advisors produced a several scenarios for the government on how NAV CANADA might arrange financing. The Department of Finance also provided significant advice on this transaction.

It was possible that NAV CANADA would not be able to raise the full sale price and obtain start-up capital at the transfer date – the revenue stream had not yet been established. Government would then need to lend money to NAV CANADA by deferring payment on a portion of the sale price. In this case, government debt would have to rank equally with any other debt of the corporation.

NAV CANADA had no equity to cushion the debt, and would have to raise full financing using future revenue as collateral, or revenue bonds. The revenue bond market was well established in the US, but was small in Canada. Consequently, financing on US markets would be important. But how would US investors react to a start-up company, where government ownership was the norm? Good ratings would be critical from US bond rating agencies.

In the end full financing was arranged by NAV CANADA and its financial advisors. The bond rating agencies gave NAV CANADA a high rating based on the following factors:

- the statutory monopoly provided by the government in legislation;
- the highly constrained ability of air traffic to avoid Canadian controlled airspace;
- the fact that the government would provide income to NAV CANADA for a transition period of two years, approximately equivalent to air ticket tax revenues collected by government;
- a high level of insurance;
- projected user charges comparable to other OECD countries;
- a flexible system for adjusting user charges;
- good enforcement powers for non-payment (seizing aircraft);
- the support provided by users of ATC, the government, the public, the employees and international air transportation agencies which indicated that NAV CANADA would run smoothly.

NAV CANADA initially secured a loan from a consortium of banks for about $3 billion, half to purchase the system from the government, and half to provide working capital and to establish reserves. NAV CANADA had no problem attracting bank support – the call for participation was heavily over-subscribed. Consequently, the government received payment in full from NAV CANADA, on the sale date, of $1.5 billion.

Shortly after the sale, NAV CANADA retired a large portion of the bank debt by issuing revenue bonds. Again, there was no problem attracting investors.

Insurance

During the consultation phase in 1994, the government worked with a large insurance company overseas that specialised in aviation insurance. The objective was to get an estimate of liability insurance costs for the most catastrophic situation that could be envisaged – two large jetliners colliding over a busy urban area. It was necessary to provide the insurers historical information on the safety record of the Canadian air traffic control system, and details of current safety practices. At that time, the insurers estimated that the new corporation should carry about $1billion US liability coverage at a premium of about $1 million per year.

NAV CANADA obtained quotations from several insurers, and had the policy in place before the closing date. The government required, through an agreement negotiated between Transport Canada and NAV CANADA, that NAV CANADA indemnify the government and carry aviation operations liability insurance of at least $1 billion per occurrence. The government also required that it be named as a co-insured in the policy.

Economic Regulation

The Canadian airlines and general aviation users were particularly concerned that the new organisation would be operated with extensive monopoly powers, and that as captive clients, they could end up paying monopoly prices. 'Gold-plating' was another fear. The government studied alternatives with assistance of a firm specialising in economics.

The findings were that NAV CANADA would operate differently than other utilities, such as electricity and telephone, where their users tended to comprise many individuals without organisational strength. NAV CANADA's users, on the other hand, would be well organised, represented by associations and able to influence price setting.

In addition, the non-share capital form of corporation meant that there would be no shareholders, and no means to distribute excess revenues other than through reducing user fees or retaining these for re-investment, reserves etc. Also, the commercial and business users were members of the corporation and appointed directors who would be sensitive to the air carrier needs and who should encourage cost control efforts within the new corporation. Furthermore, international agreements required that the air navigation system recover total costs only. Finally, debt holders would look to NAV CANADA to maintain smooth relations with its users to keep its high debt rating; otherwise, debt service costs would increase as debt was down rated.

Consequently, a system of light economic regulation was designed whereby the corporation would be required to adhere to certain legislated principles when setting prices, and whereby users of the ATC services could appeal price changes that did not appear to follow these principles to the Canadian Transportation Agency. There would be no pre-approval by a government body of price changes by the corporation, in other words, there was established a system of self-regulation.

The government did provide a mechanism for Ministerial approval of charges without appeal rights during the first two years of operation of NAV CANADA. The Canadian government was not aware of a similar economic regulatory regime used elsewhere and pioneered this approach in the federal government.

The principles included:

- a transparent methodology, that may recognise both cost and value of service;
- a price structure that does not impede safe practices (e.g., no individual charges for weather briefings);
- no differentiation between international and domestic flights, between foreign and domestic carriers, or between northern/remote services and services provided elsewhere in Canada (with cross-subsidies if necessary);
- separate charges for en-route and terminal services;
- charges for recreational aircraft that were not unreasonable or undue;
- charges consistent with international obligations of Canada;
- total revenues not greater than the corporation's financial requirements in relation to the provision of civil air navigation services.

There were other requirements put into the legislation that, for example, specified notification obligations and appeal procedures. As of February 2003, this system has worked very well with no appeals lodged with the Canadian Transportation Agency.

Safety Regulation

Transport Canada developed new safety regulations for air navigation to deal with the separation of regulator and service provider. It was a condition of closing that these regulations had to be in force. The new regulations provided that NAV CANADA would provide the same services and meet the same safety standards on day one of operation as Transport Canada had provided the day before.

Safety officials could review subsequent proposed changes to either services or procedures. These officials could require NAV CANADA to provide an analysis that detailed the impact on safety of the proposed changes. If the government did not object within a fixed time period of receiving the analysis, either on grounds of the quality of the analysis or on the impact estimated, NAV CANADA could proceed with the change.

The air navigation regulations are detailed under part VIII of the Canadian Aviation Regulations. The principles of these are:

- minimum regulatory intervention;
- risk analysis approach;
- cost-effectiveness;
- compliance with ICAO requirements;
- incorporation of departmental internal standards.

The introduction of the regulations for ATC signaled a shift toward a performance-based regulatory regime, whereby the Canadian government specifies what safety performance is required without prescribing how (e.g., procedures) this is to be attained.

National Defence

The Department of National Defence entered into negotiations with NAV CANADA and reached agreement before the closing date. The Department had accepted the privatisation scenario during consultations, and subsequently worked with the corporation as it did with other commercial providers of services to the military.

The two parties agreed that the military could make use of NAV CANADA services at no charge, and that as a *quid pro quo*, NAV CANADA could charge commercial air carriers its standard rates when they flew in airspace controlled by National Defence. This was stipulated in the legislation.

A formal agreement between NAV CANADA and National Defence specified details of the the exchange of information and cooperation on technical matters, and confirmed that NAV CANADA would assume the rights and obligations of the Minister of Transport contained in the many written and verbal agreements between Transport and National Defence. These agreements would later be replaced by legal arrangements. NAV CANADA was also restricted from charging foreign state and military aircraft for the use of its services, unless the government gave permission.

Powers of Direction

Legislation was drawn up that specified the powers of the federal government to direct NAV CANADA to provide certain services, and whether the corporation would be compensated for that direction. The direction powers are:

- services at northern or remote locations; with compensation;
- services required by international agreement; no compensation;
- services in the interest of national security specified by the Minister of National Defence; with compensation.

The federal government may also specify services required for safety purposes under the Canadian Aviation Regulations, without compensation.

Protecting Employees

Labour unions represented about 90% of the employees that would be affected by the transition to a private corporation. The commercialisation of the ANS could not proceed smoothly, unless the unions were convinced that this was in the best interests of their members. Otherwise, there would be protests, public appeals, and possibly work disruptions.

In addition to unionised staff, the employment rights of managers and non-represented employees required.

All understood and supported the idea that employees would be subject to normal provisions of the Canada Labour Code affecting commercial entities rather than the Public Service Employment Act.

Issues that were vital to employees and the unions included: wages, benefits, pensions, job security, successor rights (unions would continue to represent employees), severance pay, recognition of prior service and carry-over of accumulated leave entitlements. These obligations were guaranteed by an agreement entered into by NAV CANADA, Transport Canada and the nine bargaining agents, for a transition period between the closing date and, for each bargaining unit, the execution of a new collective agreement with NAV CANADA.

In addition, to ensure that the new organisation was responsive to employee needs, the unions argued for several seats on the Board of Directors. It was agreed in the by-laws of the corporation that the Board of Directors would include two seats to be nominated by the organisation's unions.

It was further agreed that all employees working in support of the air navigation system would be guaranteed continuity of employment during the transition. Transport Canada determined the names of employees that either worked directly for air traffic control or provided support services in such areas as personnel and finance. The Minister approved this list and provided it to NAV CANADA. The department laid off each employee from the public service, but these employees would not be entitled to any job security provisions of the government (e.g., buy-outs). NAV CANADA would simultaneously make an offer of employment to these same employees in the same area of work and with the equivalent salary and benefits. The employees did receive severance pay as would any public servant laid off from the government.

The corporation could subsequently manage its workforce as any other private employer, subject to the Canada Labour Code, union agreements and the job security provisions specified below.

Job Security

In privatising nationally owned airports, Transport Canada had required the new private operators to offer employment to all federal employees at a particular airport, and to guarantee no lay-offs for two years. However, the air carriers felt that two years of guaranteed employment could restrict the commercial freedom of the new ATC entity to achieve efficiencies in the early years. Also, some employees were concerned that after the two years had expired, they could be dismissed without protection.

The government proposed that the temporary provisions protecting employees during the downsizing of government, an active initiative at the time, be offered by the new corporation to the transferred employees. The downsizing protection program consisted of two elements in the event of lay-off: a provision for early retirement without penalty if age 50 or older; or a financial separation package prescribed by formula. The laid off employee could choose one of these two options.

These provisions would provide financial protection for employees laid off and at the same time would allow the new corporation to make workforce adjustments as necessary.

The employees, if they had stayed in the Department of Transport, did not have an employment guarantee, given the significant downsizing that was taking place in government. The previous Work Force Adjustment package ensured that a laid off employee would receive a reasonable job offer elsewhere in the government; however, this was not considered to be feasible for those departments most affected by downsizing. This included Transport Canada. Instead, the employees received the same protection package from the new corporation as they were then entitled to from the department. The unions and NAV CANADA accepted these job security arrangements.

The Right to Strike

The most powerful labour union in the air navigation system was that of the air traffic controllers. Under government control of the system on grounds of national interest, air traffic controller's right to strike had been effectively removed. Under a private corporation, however, they would fall under the same labour code that governed other private corporations, and without special provisions in legislation, they would regain the right to strike. The possibility that air service could be shut down nationally through strike action by private sector employees was a serious concern.

The Cabinet and Parliament made the decision that returned the right to strike for air traffic controllers. Many strongly believed that the new entity and its employees should be truly commercialised and should enjoy the same freedoms to resolve collective bargaining issues as other private sector firms. If the government felt strongly that a labour dispute in the air traffic control system was seriously affecting the national economy, the controllers could be legislated back to work with an imposed dispute resolution mechanism. This had been done in several cases in other economic sectors such as grain handling and shipping.

The federal government decided that the right to strike under commercial law would not be restricted by legislation, provided that emergency and humanitarian flights, such as those involved in search and rescue, and air ambulance services, continued to receive air traffic control services during a work stoppage. In addition, both the new corporation and the unions were required to ensure that National Defence personnel were able to continue to provide air traffic control services for military purposes in the event of a strike. These conditions were imposed in the legislation.

Legislation

Special legislation governing the creation and operation of the air traffic control corporation was passed by Parliament in June 1996. This Civil Air Navigation Services Commercialization Act provided the authority for the privatisation. The key components of the enactment include:

- definitions of civil air navigation services;
- the transfer of assets;
- the transfer of employees;
- the continuation of collective agreements;
- the granting of certain powers to NAV CANADA (e.g., monopoly), and the imposition of certain obligations;
- the right of NAV CANADA to introduce user charges for both the availability and the provision of services;
- the establishment of an economic regulatory framework;
- the elimination of the air transportation tax in two years, and the provision of transition payments to NAV CANADA for the same period for a fixed amount, equivalent to anticipated revenues from the tax; (There was a provision in a separate agreement for recommending to Parliament that it reduce this fixed amount, and the air ticket tax, based on expected revenues from any new user fees introduced by NAV CANADA during this period.)
- a public process for changing services and closing facilities;
- preservation of air navigation services to northern and remote communities, including a special process involving provincial and territorial governments for service reductions in these areas proposed by NAV CANADA;
- provisions for continued provision of humanitarian and emergency flights during labour disruptions, including police and military flights.

Overall the legislation proved to be complex and lengthy, and involved interested parties in its development (e.g., other departments, NAV CANADA, labour unions). About one-third of the Act was devoted to human resources and labour relations. In addition, there were control and enforcement procedures established that restricted the corporation from amending its letters patent and certain by-laws without permission of the Minister of Transport. These procedures also allowed for public appeal to the courts for alleged breaches of certain provisions of the Act, and provided for fines to be imposed when violations were proved.

Communications

The importance of open and inclusive communications cannot be over-stressed. This was especially true in the case of communications with employees. During negotiations, Transport Canada provided one employee to the negotiating team whose sole purpose was to focus on communications. A newsletter was produced regularly that discussed progress of the negotiations, answered employee questions and provided pertinent information. Sessions were held with employees across the country during both the consultation and negotiation phases where managers and members of the commercialisation team presented current developments, and fielded questions from the audience.

Post Transfer

The corporation purchased the right and obligation to provide air traffic services in Canada as of November 1, 1996. This followed payment to the government the previous day of $1.5 billion in cash and letters of offer to all employees of the government associated with the provision of these services.

The numerous undertakings, particularly as concern ANS facilities on airports and lands owned or operated by third parties, continued for several years after the transfer. Significant progress was made to secure new agreements for NAV CANADA'S occupation of these lands, but it was difficult to establish priority with these third parties, particularly other governments. In some cases, financial compensation to these third parties was necessary to complete the transactions.

The environmental arrangements for land transfers were expensive and lengthy (there were over 1400 land sites). In each case, environmental studies were required leading to remedial actions where necessary.

Labour negotiations with the air traffic controllers' union took place shortly after the transfer date. An offer by the corporation of a pay raise of greater than 30% over three years was rejected by the membership, apparently because of working conditions modified in the offer. Subsequent negotiations took place up to the strike deadline. The government stated that it would use legislation to end a strike if necessary. A deal was established before the deadline.

As of the date of writing this chapter, NAV CANADA has operated quietly and effectively, reducing internal costs and providing efficient service to users of the system. It consists of 7 Area Control Centres, one stand-alone terminal control unit, 78 Flight Services Stations, 41 radar sites, and about 1,400 ground-based navigational regional aids. In 2003 it was employing nearly 5,500 people.

User charges initially declined but were subsequently increased slightly (about 6%) in 2002 to offset reduced revenues following the events of September 11, 2001, although this was accompanied by the introduction of a deferred payments scheme. It has reduced costs to compensate for the loss in revenue, minimised its financial losses through use of its rate stabilisation fund, and has maintained its high bond rating. The result is that in 2003 its revenues were $669.2 million (compared to $592.7 million in 2001).

There have been no complaints lodged with the Canadian Transportation Agency on user charges. Users appear satisfied, even the general aviation sector that now pays a modest annual flat fee by weight of aircraft for unlimited use of NAV CANADA services. NAV CANADA continues to operate safely. A joint Transport Canada/NAV CANADA Safety Oversight Committee focuses attention on key safety management issues. The government monitors the process, regulatory compliance and activities of the corporation for safety, and is satisfied that NAV CANADA is operating in a safe manner.

Conclusion

Although many other countries, including the US (Executive Oversight Committee, 1994) looked at the privatisation of their air navigation systems in the early 1990s, only Canada has carried through comprehensive reform. Subsequently a number of other countries, such as the UK, have sought to separate ownership and regulation of their systems although in a variety of different ways. In the UK's case this involved in 2001 the selling of 46% of the NATS system to the Airline Group (made up of the UK's main airlines), giving 5% to employees, with the government retaining 49%. Other countries have different formulas, but as of 2003 there were 29 user-fee-supported ATC 'corporations' in the world, including Germany, South Africa, Switzerland, Australia, New Zealand and the UK, as well as Canada, that accounted for about 80% of global air traffic (Poole, 2003).

The privatisation of the Air Navigation System in Canada has been a successful venture. Many stakeholders were involved in helping guide the policy that led to the creation of the new corporation and in working together to help form the consensus necessary to move the air navigation system smoothly out of government. Although the negotiations were complex and the implementation arduous, the transition from a departmental function that was intertwined with government operations, to a stand-alone private, corporation occurred in minimal time.

An alignment of forces and pressures and the commitment of the Minister of Transport influenced the transition. In the case of air traffic control, government is now truly steering, while the private sector does the rowing. But it should be noted that the direction in which the government is steering was not arrived at alone. Open and inclusive consultation is at the heart of the Canadian success.

References

Dey, P. (1994) *Where were the Directors? Guidelines for Improved Corporate Governance in Canada*, Toronto Stock Exchange, Toronto.

Executive Oversight Committee (1994) *Air Traffic Control Corporation Study*, Office of the Secretary, US Department of Transportation, Washington, DC.

Poole, R.W. and Butler, V. (1996) *Reinventing Air Traffic Control: A New Blueprint for a Better System*, Policy Study No. 206, Reason Foundation, Los Angeles.

Poole, R.W. (2003) *Air Traffic Control Systems Weather the Storm*, Reason Foundation, Los Angeles.

Transport Canada (1994a) *Safety Regulation*, The Study of the Commercialization of the Air Navigation System in Canada, Discussion Paper 2, Ottawa.

Transport Canada (1994b) *The Need for Economic Regulation of a Commercial Air Navigation Organization*, The Study of the Commercialization of the Air Navigation System in Canada, Discussion Paper 3, Ottawa.

Transport Canada (1994c) *International Experience of ANS Commercialization,* The Study of the Commercialization of the Air Navigation System in Canada, Discussion Paper 4, Ottawa.

Transport Canada (1994d) *Illustrative Charges*, The Study of the Commercialization of the Air Navigation System in Canada, Discussion Paper 5, Ottawa.

Chapter 3

The Delta Launch Vehicle: A Model of Government-Industry Cooperation

Francis T. Hoban and John Mulcahy

Introduction

One of the greatest achievements in the history of the exploration of space was the Delta Launch Vehicle program (Winter, 1990). Figure 3.1 provides an illustration of the equipment involved.

The Delta program is important for a variety of reasons not least of which was that it was a model of public-private partnership. However, very few people know the story of the unique arrangement between the National Aeronautics and Space Administration (NASA) and private industry that made Delta such a success.

The NASA Delta program began on April 28, 1959, with the signing of a $24 million contract with the Douglas Aircraft Company for the acquisition of 12 launch vehicles. Since then there have been 280 successful Delta launches with just 15 failures (through 2002), a 95% success ratio that has earned Delta the worldwide nickname of NASA's workhorse. Delta's capabilities have grown from 100 lbs of payload for geosynchronous, GTO, missions in 1959 to 4,000 lbs, with an accompanying increase in volume.

Success comes about for a variety of reasons. Some institutional, some pure luck but in some cases it is heavily influenced by the contribution of individuals. Delta's outstanding record was due in large part to the leadership brought to the program.

William Schindler, Delta program manager from 1964 to 1976 at NASA's Goddard Space Flight Center (GSFC) in Maryland, influenced Delta's history more than any other individual. Schindler was known for his devotion to customer needs, for his faith in people, and for his creativity, perseverance and imagination. He built an outstanding industry and government team that consistently delivered successful, low cost launches, meeting a wide variety of requirements for a worldwide clientele. Schindler's vision enabled the Delta team to play a foundational role in the early years of the space communications industry. Bill Schindler left NASA in 1976. After his departure from NASA, the Delta Program he led for 12 years in the 1960s and 1970s went on to achieve many new successes, including the introduction of the Delta IV Launch System in November 2002.

This study describes the events surrounding one of the achievements of the Delta effort, the Delta upgrade known as '3914.' (The four digit designation system describes the rocket: the first digit represents the first stage, the second digit is the number of augmentation motions; the third digit represents the type of second stage, and the fourth digit, the type of fourth stage.) According to John Christopher, Vice-President of Space Programs at GE American Communications, Inc., the Delta upgrade was essential to the successful development of the satellite communications industry: 'Bill Schindler's aggressive marketing of Delta upgrades led to the RCA/McDonnell Douglas 3914 Program, the first true commercial launch vehicle hardware venture. That contribution to RCA's Satcom Domestic Satellite Program was pivotal, and thus he helped start a major industry.' (Lynch, 1998) The successful outcome of this upgrade was remarkable because it was made without the support, encouragement or even endorsement of top management at NASA.

(*Source*: NASA History Office)

Figure 3.1. Delta Rocket Carrying Mars Pathfinder

The NASA Delta team's contributions to the successful management and performance of the Delta Launch Vehicle cannot be overstated. In addition to the critical role the Delta has played in the establishment of the satellite communications industry, the technical success of Delta has also contributed greatly to the achievements of space science, and to advances in weather/hurricane predictions through satellite technology.

But perhaps the Delta team's greatest legacy was its almost unprecedented success in bringing together government, launch vehicle contractors, users, and commercial customers to work as a team for the advancement of aerospace and space commercialization. Without this cooperation, the development of the satellite communications industry especially would have been seriously delayed.

Background

In 1954, at the height of the Cold War, the US Air Force issued a Request for Proposal for a rocket that could carry a nuclear warhead about 1,500 miles, roughly the distance from Northern England to Moscow (Launius and Jenkins, 2002). The Douglas Aircraft Company of Long Beach, California responded to the RFP with a design for the Thor missile.

The Thor was designed to use the only engine readily available (from the previously developed short-term Navajo cruise missile), and a guidance system from the Atlas missile. Only seven months after winning the contract, the first 65-ft Thor was ready to be airlifted to Europe by a four-engine Douglas C-124 military transport. By the end of 1958 several more Thors had been deployed to England, where they remained until replaced by more powerful rockets in August 1963.

In 1957 the Russians launched the first satellite to orbit the earth, Sputnik. The United States responded by establishing the National Aeronautics and Space Administration (NASA) as a successor to the National Advisory Committee for Aeronautics (NACA), a civilian space agency, to counter what the Eisenhower Administration viewed as a threat to national security in space. In April 1959, NASA asked Douglas Aircraft to modify the Thor for civilian use. The Thor had been modified for two missions already, once with the second stage of the Vanguard rocket for a US Navy program, and also with a solid third-stage motor that later flew in the Pioneer program.

The third modification of the Thor, this time for the newly created NASA, also required the use of an assortment of existing parts in order to be ready by the deadline of 18 months. A Rocketdyne MB-3 engine fueled by kerosene and liquid oxygen and using upper stages from the Vanguard launch vehicle, powered the Thor first-stage booster. An engine from the Aerojet AJ-10 using nitric acid and hydrazine powered the second stage. The third stage was a solid motor from the Allegheny Ballistics Laboratory, the X-248, with upper stages from the Vanguard launcher. This new launch vehicle was called Delta, since it was the fourth generation Thor.

The development of the Vanguard 2nd and 3rd stages and the upgrading of the Viking 1st stage had been funded by the Navy for non-military purposes before the creation of NASA. The development, certification and manufacturing costs were borne by the military in the 1950s. Thor-Delta, the world's first major civilian launch vehicle, incurred just a few incremental manufacturing costs from its military progenitor. Thus, NASA got its first real bargain.

This relatively quick, cheap assembly of the best available existing components failed during Delta's maiden launch in May 1960. NASA engineers quickly identified and corrected three possible causes, any one of which would have resulted in mission failure.

Three months later, Thor-Delta was launched again, carrying Echo-IA, a huge (100-ft diameter) aluminized Mylar balloon, developed by William Sullivan at the NASA Langley Research Center. When inflated in orbit, the Echo-IA bounced two-way voice signals between east and west coast ground stations. This passive system constituted the first communications satellite in history. Subsequent telecommunications flights solidified Delta's reputation as the 'workhorse of the revolution in wireless communications' (Forsyth, 1999).

'Delta was, from its inception, intended for a multi-orbit capability,' says Al Jones, a Goddard Space Flight Center Delta Project team member in an interview with Frank Hoban on May 17, 2000. 'There were LEO missions planned, as well as highly eccentric scientific orbit requirement...When (the) Hughes (Corporation) conceived the synchronous spinner for use on the Scout launch vehicle, NASA management felt more weight margin was needed and put it instead on a Delta, along with the first strap-on rocket boosters.'

By the mid-1960s, strap-ons for the Delta, solid rocket Castor boosters, could send 1,593 lbs to LEO, low earth orbit, or 1,060 lbs. to GTO, geostationary orbit. Delta's first-stage liquid propellant engine was upgraded from 172 K lbs. to 205K lbs. of thrust in the mid-1960s when higher thrust (higher payload) Saturn I-B engines were surplused by NASA for use on the Delta program, another low-cost upgrade.

In 1967, Douglas Aircraft became incorporated into the McDonnell-Douglas Astronautics Company (MDAC). Soon the performance of Delta was enhanced with a second stage storable propellant engine from TRW – Thiokol Corporation provided an enhanced performance third stage motor originally developed for NASA's Lunar Survey mission, thus incurring no development costs. But by 1969, with the Delta 2914 model, the program had run out of cheap upgrades. In order to carry bigger payloads into space, Delta would need a costly, major upgrade. However, Delta would have to compete for funding with the Space Shuttle, NASA's next big program after Apollo.

Following the Moon Landings

In the 1960s, Delta suffered only six failures out of 74 launches; two of these occurred nearly back to back in July and August of 1969, right after Neil Armstrong and Buzz Aldrin touched down on the moon. Nevertheless, Delta remained the expendable launch vehicle (ELV) of choice, selected for about half of all NASA launches.

With seven successful launches, 1970 was a good year for Delta (see Table 3.1 for a brief account of subsequent development). However, in 1971 two more anomalies occurred. On September 29, 1971 Delta 85 successfully carried the Orbiting Solar Observatory-7 (OSO-7) into orbit from Cape Canaveral. However, the spacecraft suffered a nitrogen gas leak in the second stage hydraulic reservoir, directing the OSO-7 away from its planned orbit. Although some managers believed that the science return from the 'wrong' orbit was better than expected from the 'planned' orbit, the leak was a problem.

Delta 86 was launched on October 21, 1971, from the Western Test Range in California, carrying the ITOS-B spacecraft. The rocket veered off course, probably from a leak in the Delta oxidizer system's vent valve. The attitude thrusters apparently fired continuously until their nitrogen fuel was depleted, plunging the rocket and payload into the Arctic Ocean.

Table 3.1. Chronology of Thor-Delta Development and Operations

Date	Event
June 26, 1970	NASA awarded McDonnell Douglas a contract to incorporate the new Delta inertial guidance system into the Thor-Delta vehicle.
October 13, 1971	Details of the improved 2000 series Thor-Delta were discussed with representatives of potential user organizations.
March 11, 1972	The first Thor-Delta with a Universal Boat Tail was launched (TD-IA); the boat tail allowed the addition of up to nine strap-on solid rocket motors.
July 23, 1972	The launch of Landsat I marked the first use of nine strap-one and the new uprated second-stage engine (AJ 10-118F). This Thor-Delta model was designated the 904.
September 22, 1972	With the launch of Explorer 47, the first 1000 series Thor-Delta was proven successful.
November 15, 1974	For the first time, a Thor-Delta launched three satellites simultaneously (NOAA 4, Intesat, and Amsat Oscar 7).
May 7, 1975	The launch of Anik 3 marked the 100th successful Delta liftoff.
Spring 1976	NASA officials studied the possibility of using a Delta-class (TE 364) stage instead of the Air Force-sponsored Interim Upper Stage for use with Shuttle payloads.
May 1976	The U.S. Aeronautics and Astronautics Control Board, made up of NASA and Department of Defense personnel, approved the Delta 3914 model for government use.
November 23, 1976	NASA Headquarters plans called for Delta to be phased out at the Kennedy Space Center during 1980 in anticipation of the Space Transportation System becoming operational.

In response to these two failures, John Naugle, NASA's Associate Administrator for Space Science and Applications, appointed a Delta Launch Vehicle System Review Board, as required by NASA Policy Directive (NPD) 8621.1A, effective November 4, 1970 (NASA News Release, 1971). The NPD stipulated that, 'with respect to all accidents and mission failures, NASA will:

- take immediately all possible actions to prevent injuries and losses,
- conduct investigations to determine the actual or probable cause(s) as a basis, for avoiding recurrence,

- take preventive measures and any other appropriate actions to avoid recurrence.'

The NPD made NASA Headquarters program and institutional directors responsible for 'ensuring that plans and other arrangements exist to cope with all accidents and mission failures within their respective jurisdictions, to investigate them, and to report the results thereof to authorities concerned' (NASA Policy Directive 8621, November 4, 1970).

William Lucas, Deputy Director of Marshall Space Flight Center (MSFC) in Huntsville, Alabama, headed the Delta Review Board. Its task was to investigate the ITOS-B failure and other Delta problems. The authorities and responsibilities of the Review Board included directives to determine the actual or probable causes of the accident or failure, to develop recommendations for preventive or appropriate actions, and to provide a detailed written report to the Administrator or Deputy. The point man for the Delta Review probe at GSFC, Bill Schindler, had a much-deserved reputation for achieving mission success at low cost. He often flew improvements without benefit of test flights. Schindler took other risks, but was never reckless. Although some disagreed with his management style he was the epitome of the 'new' low-cost NASA manager.

The Delta Project Manager

Bill Schindler was born August 10, 1927, in Wisconsin. He served in the US Navy from 1945 to 1948 and again during the Korean War from 1951 to 1952. He graduated from the University of Maryland in 1957 with a B.S. in mathematics. He worked with the Vanguard Project in Florida, but left in 1959 to work on the Delta Project at the GSFC in Greenbelt, Maryland. He was appointed Delta Project Manager in 1962, a post he held for the next 12 years. He led a project team of more than 1,000 skilled government and industry personnel, a team that built and launched a hundred Delta rockets. 'Bill Schindler was the son of a steelworker, a pragmatic, down-to-earth person,' recalls Pete Eaton, another Delta team member, 'sharp, alert, incredibly bright, absolutely fearless.' (Interview with F. Hoban, June 26, 2000).

A commemorative magazine honoring Schindler, issued to mourners upon his death by colleague Jim Lynch, describes Schindler's management genius:

> Unlike many executives, the words 'bureaucracy' and 'committee' were not in his lexicon, enabling him to cut through institutional red tape to get the job done. His close associates remember him for his creativity, honesty, competitiveness, and indefatigable appetite for work, all fueled by an unparalleled wit and magnetic personality. Bill Schindler had a vision and a simple philosophy – keep costs down, make it work, and always care for and satisfy the needs of the customer. This logic filtered down to every member of the NASA/McDonnell-Douglas team, bonding a lifetime partnership (Lynch, 1989).

Schindler's accomplishments earned him high NASA honors, including the Distinguished Service Award. The National Space Club honored him in 1980 with its prestigious

Astronautics Engineer Award, and in 1991 with a Special Achievement Award 'for management of the evolutionary development of the Delta Program' (Lynch, 1989).

'Bill Schindler was the most dedicated and creative person I have ever met,' says C.A. Ordahl, Vice-President of the Advanced Product Development and Technology Division at McDonnell Douglas. 'His management philosophy was totally focused on what was best for the Delta Program. He demonstrated that a small management team, with simple but reliable vehicle upgrades, and low cost, would sustain the vehicle for years into the future. His legacy remains in the Delta of today and tomorrow' (Lynch, 1989).

Findings of the Failure Review Board

The Delta Failure Review Board headed by Lucas reported their Delta findings on December 17, 1971, calling for a costly 90% rate of reliability. In the transmittal memorandum to Naugle (December 17, 1971), Lucas states:

> While recognizing the very significant achievements of the Delta Program in the last several years, the Board believes that the reliability of the Delta can be improved through increased rigor in engineering, manufacturing, checkout, and quality control. The Board recognizes also that reliability improvement will increase costs, but this increase may be fully justified by the important and increasingly expensive payloads that Delta is being called upon to orbit for foreign nations, other Federal agencies, and the private sector.

The Board concluded that the Delta program at Goddard was driven by 'underlying, but unwritten, philosophies, which have stressed cost minimization and improvements through experience,' or corrective actions after accidents happen. The failure board concluded:

> With proper management within the Government and at the contractor's facilities, the additional costs for this goal [of 90% reliability] should not be exorbitant. To increase the reliability of the system beyond 90%, considering the age of the design and the attendant manufacturing techniques, the costs may be prohibitive, but such a determination was beyond the purview of this board.

A concise summary of the Failure Review Board findings was provided by Richard Smith, manager of the Saturn program at the Marshall Space Flight Center (MSFC). Smith conducted a similar review of the Thor and Delta failures '...to avoid the problems that befell the Thor-Delta Program (Memo from R.G. Smith, MFSC, to PM-Dir/Mr. Shepherd, MSFC, January 25, 1972).' Smith asserted that the Delta launch vehicle had a history of 'ground and flight leaks derived from poor design, lack of discipline in manufacturing and quality control, and inadequate test requirements and systems verification procedures.' In addition, he said, the Delta Failure Review Board found:

- The Delta system was subject to a number of contamination sources that 'clearly indicate a lack of control.'

- The Delta project did not have a spares program, 'resulting in cannibalization practices.'
- Reworked hardware was reinstalled on a flight vehicle without positive isolation or corrections.
- There was 'evidence of a lack of discipline in the management approach,' with little Government technical penetration of the program.
- There was a lack of adequate working relationships between Goddard Space Flight Center and Kennedy Space Center.

The last point refers to the deterioration of the working relationship between the two centers after Kennedy Space Center lost work in payload integration to MDAC. Schindler had transferred this work to MDAC because they could do it more cheaply. With the results of the Failure Review Board in, Schindler and the entire NASA project manager community got the *real* message: 'NASA will accept no failures. We will forgive you if you overrun costs, but we won't forgive you if you fail.'

Schindler and Delta survived the Review Board. However, the Board's findings, including the estimated cost of at least $30 million to correct the deficiencies, gave NASA Headquarters the grounds to deny any further upgrades to Delta or to any other NASA expendable launch vehicle, except for the much smaller Scout. The $30 million needed to enable Delta to reach 90% reliability would have to come out of a NASA budget centered on NASA's next big leap into space, the Space Shuttle.

Delta flights resumed on January 31, 1972, completing seven successful flights during the year. The Shuttle was far from being designed, much less in production. Even the Lucas Report acknowledged the short-term need to rely on Delta:

> While long-range plans call for phasing Delta out when the Space Shuttle becomes available in the late 1970s, analysis of firm and highly probable future missions indicates that some 40 to 60 Deltas will be required before the Shuttle becomes operational (Lucas, 1971).

After the Failure Review Board

For NASA, the early part of the 1970s was a time of both great triumph and great tribulation. Apollo was an overwhelming success. There was a widespread belief that NASA could do anything. Yet the government and public that praised the moon landings so highly had nothing more for NASA to do. The NASA Deputy Administrator who had been part father of this young agency, George Low, now had to become its reinventor.

George Wilhelm Low was born on June 19, 1926, near Vienna, Austria. He studied aeronautical engineering at Rensselaer Polytechnic Institute (RPI). In 1950 Low joined the National Advisory Committee for Aeronautics (NACA) as an aeronautical research scientist at the Lewis Flight Propulsion Laboratory in Cleveland, Ohio. In 1958 he worked on the planning for the new agency, NASA, and for Project Mercury. Soon afterwards he transferred to NASA Headquarters as the chief of Manned Space Flight, where he became Deputy Associate Administrator in 1963. In 1964 he became the Deputy Director of the Manned Spacecraft Center in Houston, Texas. As Apollo Spacecraft Manager, he directed eight successful Apollo mission. He won the admiration and respect of NASA and industry colleagues alike for his technical competence and his managerial abilities.

In December 1969, George Low was appointed NASA Deputy Administrator. These were trying times for the agency. The Kennedy Administration's mandate to land astronauts on the moon and return them safely had been met, and NASA had completed Skylab. But NASA did not have another large program on the drawing board and was an easy target for cost-cutting as Congress considered price controls and budget cuts. Historically, it was a time of runaway inflation, anti-war protests, riots in the streets and political turmoil. The space race was giving way to economic concerns, and space endeavors did not seem to fit in the new political world order.

Low needed a development program large enough to maintain the NASA Apollo infrastructure. The new program must involve multiple NASA Centers and contractors, with money flowing to as many Congressional districts as possible. That program was the Space Shuttle. Just four months into his term as Deputy Administrator at NASA Headquarters, Low began to sell the Space Shuttle to the Administration, the Congress, a dubious science community, and, not least, to some in NASA. In 1970, a joint Department of Defense (DOD) and NASA study team assured President Nixon that a fully reusable Space Shuttle system (capable of meeting both NASA and DOD requirements) could be developed within eight years for $5 billion. It would be even easier and cheaper without the DOD requirements.

The study estimated that each orbiter could be built for $250 million, with a per-flight cost of just $5 million. At that rate, the Shuttle's 50,000 lbs payload to low Earth orbit would cost a mere $100 per pound (Science and Technology Panel of the President's Science Advisory Committee, 1970). However, to achieve that launch cost it would be necessary to put *all* payloads on the Space Shuttle. Low decided to proceed with the Space Shuttle, even though some believed that the per pound payload estimate could be off by a factor of 10. He argued that significant savings would be realized if the Space Shuttle could capture the total payload market and make weekly trips to outer space. Obtaining federal funding for this capability would occupy most of George Low's time and effort.

Industry Pressure to Upgrade

With the Shuttle years away, industry still needed low-cost, reliable space transportation, and in most cases that meant Delta. Wernher von Braun, then vice-president of research and development at Fairchild Hiller, wanted to use the Delta to launch a Fairchild spacecraft. He pressed the Deputy Administrator for an upgrade. George Low met with him in October 1972. Von Braun, according to Low, argued that using an Atlas Centaur would use up all of Fairchild's profits, and asked instead for a $6 or $7 million Delta upgrade to a 2,000 lbs injection capability by McDonnell Douglas. Low refused.

Low continued his discussions with Fairchild Hiller, however, suggesting that commercial firms might want to deal directly with MDAC in the effort to upgrade the Delta. The difficulties, of course, were that such an arrangement had never been achieved before, and such partnerships were not at all common. However, Low also indicated that if Fairchild Hiller could come to an agreement with MDAC, NASA might provide some consulting services by its staff at GSFC.

On November 12, 1972, a month after meeting with Wernher von Braun, the NASA Deputy Administrator met with two RCA vice-presidents. The RCA representatives said they would pay MDAC the $25 to $30 million needed to upgrade the Delta if NASA would agree to include the up-rated vehicle in the NASA launch vehicle stable. RCA was developing the 24-transponder Satcom communications satellite, and proposed that NASA upgrade the Castor II solid booster strap-ons to Castor IV's. RCA would fund development and certification.

Without this upgrade, RCA, like Fairchild, would be forced to use a much larger and expensive launch vehicle, threatening the economic viability of the project. In exchange, if the up-rated Delta launch vehicle worked, and with no cost to the Government, RCA asked that NASA make it available for future use.

This request to Low had its genesis several months earlier when RCA first proposed the up-grade to Schindler. Schindler was sympathetic to RCA's appeal and wanted to accommodate a valued Delta customer, but he had to reiterate Low's 'no up-grade' policy, noting that he could not do the project without approval from Low. RCA told Schindler that they had considered going directly to McDonnell-Douglas for the upgrade, but they needed the NASA Delta team to manage the program, and for NASA to put the new launch vehicle in the official NASA/DOD stable. By the end of the meeting with Schindler, RCA decided to take the next step and discuss the proposed upgrade with MDAC.

McDonnell-Douglas was interested in upgrading the Delta, especially with someone else paying the bill. MDAC believed it could upgrade three vehicles for about $25 million instead of the $30 million estimated by the Lucas Review Team. The company also agreed on the need for Schindler's team to manage the project. RCA went to see Low, and received the expected initial answer. But once Low suggested that RCA pay for whatever degree of NASA participation was necessary if the company elected to proceed with the upgrade directly with McDonnell Douglas, they knew they had won.

From the very beginning RCA had planned that GSFC would manage the McDonnell Douglas upgrade. A new kind of government and private industry partnership was sealed. This contractor-to-contractor arrangement, with NASA in the middle, was a novel idea. NASA held several meetings in late 1972 through mid-1973 to determine the agency's response to RCA's unusual offer. Reliability was a major concern at NASA, but RCA believed that Delta was the most reliable launch vehicle available.

On November 24, 1972, George Low drafted a NASA policy memorandum for upgrading the Delta launch vehicle, after seeking concurrence from the DOD, that also had a vested interest in the Delta. In addition, he wrote:

> If a potential commercial user requires an up-rated Thor Delta, and if the McDonnell Douglas Corporation is willing to up-rate the vehicle, then such an up-rated vehicle can be procured provided all of the following conditions are met:

- The potential user and McDonnell Douglas are fully informed of NASA's concerns and conditions as expressed in this memorandum.
- Up-rating modifications will be made only on designated vehicles after they have been procured by the commercial user. No modifications in any way connected with the up-rating will be made on any of the 'standard' Thor Deltas. In other words, up-rating changes will be 'peculiar' changes, made only to those vehicles purchased by those commercial users who have a need for an up-rated vehicle.
- All costs related to the up-rating will be borne by the commercial user requiring the up-rating. The added cost of up-rating (over the cost of a standard vehicle) will be handled directly between McDonnell Douglas and the commercial user, either on a one-time basis, or on a cost per flight basis, or both. Strict accounting procedures will be developed, to assure that all related costs, including the costs of changes and overruns, are fully borne by the user. In addition, the government will be fully reimbursed for the use of any government tooling, equipment or facilities used in the up-rating, for any required facility modifications, and for the services provided by government personnel or their support contractors.
- NASA will provide technical direction over the up-rating. The purposes of this direction are to assure that all required safety conditions are met; and to assure a reliability of the up-rated vehicle commensurate with the reliability of the standard vehicle, insofar as practical, on a 'best efforts' basis. The costs of any changes or overruns resulting from this technical direction, and the costs of government personnel involved in this work, will be borne by the commercial user.
- The up-rated vehicle will be a 'mission peculiar' vehicle, and not a member of the NASA/DOD family of launch vehicles. Therefore, if a 'guaranteed launch' policy is established, it will not apply to this vehicle. There is no intention to bring this vehicle into the NASA/DOD family, and therefore it should not be considered to be an available launch vehicle in discussions with other potential users.
- The Thor Delta project team has done an excellent job in getting the vehicle under control after two failures. Nevertheless, this team is still stretched rather thin. Their first responsibility must be to assure the continued performance of the existing vehicle. Therefore, all up-rating activities would only have a second priority on their available time.
- A family of launch vehicles – ranging from Scout to Titan/Centaur – exists in NASA and the DOD. All potential payloads can fit on one of the launch vehicles in this family. Since the decision not to uprate was made, it has been brought to my attention that this decision may put several of the applicants for domestic communications satellites at an economic disadvantage. This is particularly true since they assumed, at the time they made their application, that an up-rated Thor Delta would be available. Based on this information, and based on the results of several recent reviews and briefings, I have reached the following conclusions:

- The performance of the Thor Delta will not be up-rated for any NASA, DOD, or other non-commercial users.
- The DOD, through normal … channels, concurs in this up-rating.

On June 8, 1973, Low met again with industry and NASA officials about the Delta upgrades, this time including a representative from MDAC. He reviewed 'the planned principles of contracting between McDonnell Douglas and RCA, McDonnell Douglas and NASA, NASA and RCA and McDonnell Douglas and Thiokol. I am satisfied that the various contracting arrangements meet the requirements of my November 29, 1972, policy directive.' Then he explained the arrangement:

> I have reviewed with our people the various contractual arrangements between, and among, RCA, McDonnell Douglas, and NASA. These appeared to be well thought out, and, under the circumstances, are probably the best arrangements which we could make. McDonnell Douglas is going to charge RCA $1.2 million of nonrecurring costs on each of three launch vehicles which RCA would order. McDonnell Douglas further expects to charge off the remaining nonrecurring costs on the next five or six launch vehicles sold to RCA or to other users. They still expect to do this entire job for something of the order of $8 million. The contractual arrangements will meet all of my requirements as established in the November 29, 1972, memorandum on this subject. In addition, McDonnell Douglas intends to vest the rights in the up-rated Delta in NASA.

In a memorandum to NASA Administrator James Fletcher, Low noted: 'They also agreed to give title to the rights and data associated with the uprated Delta and its associated equipment to NASA at no expense. This will be done after the first successful launch, so that we might then make a decision as to whether we want to incorporate the up-rated Delta into the NASA stable of launch vehicles' (Memo from G. Low to J. Fletcher, June 11, 1973).

From the beginning of the Delta program, GSFC was the supplier of Delta launch vehicles to NASA and all other commercial customers. To fly on a Delta, it was necessary to go to GSFC. NASA placed an annual order with the vehicle manufacturer, McDonnell Douglas Corporation, for a varying number of vehicles on a cost-plus fixed fee basis. The contract amount was an estimate – the real cost was not fully known until the end of the fiscal year, when the cost of the Kennedy Space Center launch services were added to the cost of the vehicle from McDonnell Douglas and the management charges for GSFC launch services.

The Kennedy Space Center charge was determined by dividing the total yearly expense by the number of launches. Only then was the charge to the customer finalized. Over the years, the Delta Project Office, in an attempt to reduce the Kennedy Space Center charge, had assigned more and more of the preparation, processing and integration work of the launch vehicle to McDonnell Douglas. It was cheaper to do the work at the manufacturing site and ship a more launch-ready vehicle to the Center. GSFC provided payload integration support, customer unique requirements, safety, reliability and quality assurance, and overall project management. RCA wanted GSFC to provide its usual services for the upgrade.

The Outcome

After NASA, RCA and McDonnell Douglas put together the most innovative privatization partnership in aerospace history, Delta 96 failed on July 16, 1973. Subsequently, Delta had three consecutive successes in 1973 before another failure on January 19, 1974. However, with the launch of Westar-A on Delta 101 on April 13, 1974, Delta began its longest string of successes to date, successfully completing 29 launches before the failure of the Delta 130 in April 1977. Delta's reliability continued to increase, with only five failures in the 164 launches from April 1977 to January 2003.

By 1974, MDAC had replaced the Rocketdyne MB III main engines with RS-27A engines that provided some 25% more thrust for about the same size and weight. The newly upgraded launch vehicle could use up to nine solid rocket strap-on boosters. Since the RS-27 engines were surplus Saturn-IB engines, they did not cost anything to develop and build, providing Delta with another 'cheap upgrade.' The improved Delta produced 230,000 lbs of thrust, with its second stage using the TR-201 engine, a modification of the Apollo Lunar Module descent stage engine.

RCA launched SATCOM 1 onboard a Delta 3914 in 1975. It was used to transmit the movie channel HBO to cable television providers, thereby spawning a new industry. Everybody won. RCA launched a new space-based business, McDonnell Douglas had an entirely new capability to market to other customers, GSFC had an extended lease on life for the Delta launch vehicle, and none of this cost the American taxpayer a single cent.

Not satisfied with all that had been accomplished, Bill Schindler established a surcharge process to pay back RCA's original investment for the upgrade. Every follow-on customer that used the 3914 configuration paid a small surcharge (in the $20,000 range) until RCA was fully reimbursed for its development costs. This rebate continued until the early 1980s. The charge was so small in relation to the average total cost that no customer complained. The Delta upgrade initiative was completed through an innovative use of reimbursable money, a concept that became increasingly important to NASA during the era of the Space Shuttle.

'I have always considered the Delta one of the most successful projects in our space history,' wrote Sidney Metzger, assistant vice-president and chief scientist of COMSAT. '(Without a single test flight) it has grown from an 85-pound synchronous orbit capability in 1965 to a 1,000-pound capability in 1975. This growth has been accomplished in all cases by using engines for the various stages from other projects, which is in accordance with the rule to use existing hardware wherever possible.'

Metzger concludes: 'Another point which certainly contributed to Delta's success was brought out by a statement by Bill Schindler, that they never tried to shave weight to the last pound (a lesson he learned from Vanguard). Still another reason for its success has been the use of practically the same team throughout the 15 years of the Delta's existence' (S. Metzger, letter to R. Gilruth, March 11, 1975).

Schindler left NASA in 1976 after it became clear that his philosophy of a small, tight team keeping costs and changes to a minimum clashed with that of his supervisors at NASA. Peter Eaton recalls,

> Bill was always honest, upfront, and straightforward with everyone on how he felt about excess staffing and paperwork, whose only purpose seemed to be to save face when the inevitable failures occurred (10% of the time, on average). Unfortunately, he was characteristically clear about this in his presentations to the Delta Failure Review Board – despite my warnings that they were not interested in a debate on the subject, and it would probably prove counterproductive to deliberately antagonize them – which he was not afraid to do (Interview with F. Hoban, June 24, 2000).

The eventual rollout of the Space Shuttle was just a temporary setback for the Delta launch vehicle. President Reagan ordered the Delta back into production after the Space Shuttle Challenger disaster in 1986. Delta continues to be a low cost, reliable back-up to the Space Shuttle. Delta II had its first launch in February 1989 and has had a 98% success rate (see Figure 3.2).

(*Source*: NASA History Office)

Figure 3.2. A Delta II Launch Vehicle Being Prepared

The Delta III was successfully launched in 2000, after two preliminary failures, and the Delta IV in November, 2002 (see Figure 3.3).

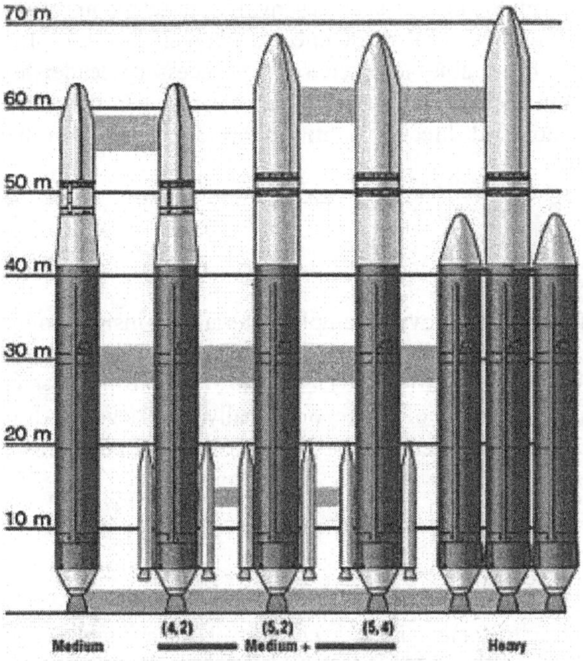

(*Source*: Boeing website)

Figure 3.3. The Delta IV Launch Vehicles

What made the success of the Delta upgrade project so unusual was industry's attitude towards government, and government's response. In an era when American corporations seek to avoid direct involvement with the government, perhaps there are lessons to be learned from a project where corporations actively sought government partnership.

Conclusions

Developing a successful launch vehicle involves overcoming the challenges of a massive set of technical, economic and management problems. Additionally, following the ideas of Napoleon, it also involves no small amount of luck. The Delta program was by any criteria successful. It also involved all of these elements. But it also embodied considerable leadership, energy, and what would in the private sector probably be called entrepreneurship on the part of a single individual, in this case William Schindler. In a sense, the success of the program represented a realization by Schindler, that the type of environment in which he was operating was not the old traditional one of a research laboratory. There were 'customers' and 'clients' to satisfy, there was a mixture of profit motivated private corporations and public interest agencies involved in development and production, and there were massive technical difficulties to overcome. While it is possible that others may have been able to exercise the necessary leadership (counterfactuals are always of intellectual interest), it seems clear that in this cases progress came about through vision and direction from the top rather than from the textbook management structures.

Acknowledgements

Special thanks go to William Lucas, who provided ample material and insight on the Delta Failure Review Board, material unavailable from any other source. Thanks also to Pete Eaton for his special insights into the heart of the Delta team, and to Bob Goss for the long hours he spent educating the writers of this report. Finally, thanks go to Alton Jones, Ike Gilliam and all the others from that era who provided their unique perspectives into this most remarkable project.

References

Forsyth, K.S. (1999) *History of the Delta Launch Vehicle*, online at <http://kevin.forsyth.net/delta/background/htm> last updated 28 July.

Launius, R.D. and Jenkins, D.J. (eds) (2002) *To Reach the Higher Frontier: A History of US Launch Vehicles*, University of Kentucky, Lexington.

Lucas, W. (1971) *Delta Launch Vehicle System Review Board Report*, Delta Review Board, Huntsville.

Lynch, J. (1989) *Delta: William R. Schindler*, Jim Lynch and Associates, Potomac.

NASA News Release (1971) 'Delta Review Board Established', October 28.

Science and Technology Panel of the President's Science Advisory Committee (1970) *The Next Decade in Space*, Science and Technology Panel of the President's Science Advisory Committee, Washington, DC.

Winter, F.H. (1990) *Rockets into Space*, Harvard University Press, Cambridge.

Chapter 4

Leasecraft

Harold Miller

Introduction

The exploration and commercialization of the Americas in the 16th century is sometimes compared to the contemporary efforts to commercialize space. Governments in search of gold and other precious metals funded early explorations of the Americas, but establishing trade and settlements in the new territories took much longer. Today the search for 'gold' includes the successful mining of radio frequency bands that permit a profitable space-based communications industry.

But the commercialization of space has had difficulty moving beyond this initial success into space industrialization. In the 1980s, one pioneering project designed to exploit space for commercial purposes failed. The factors that contributed to that failure continue to hinder the private commercialization of space.

In the US in the early 1980s, there was much enthusiasm for developing cost-effective ways to provide access to space for strictly commercial ventures. Congressional legislation authorized private enterprise to operate in space and to manufacture a wide variety of products in zero-gravity that could not be produced on Earth. The rapid success of the communication satellite business contributed to this enthusiasm.

There were very optimistic projections that the Space Shuttle, first launched in 1981, would be able to undertake weekly flights at reasonable cost by the end of the decade, making access to space a marketable commodity. Corporations envisioned commercial space factories that could produce products as diverse as ball bearings, semiconductors and pharmaceuticals. Others foresaw a space tourism industry (Tucker, 1984). National Aeronautics and Space Administration (NASA) Administrator James Beggs even contemplated selling the Space Shuttle to private interests by 1989 (*Space World*, 1984).

Several entrepreneurs and start-up ventures proposed aggressive space investments. In 1982, Fairchild Industries, Inc., of Germantown, Maryland, proposed to develop a satellite called Leasecraft that would be available to all customers for manufacturing a variety of goods in an orbiting facility. The Leasecraft would be placed in orbit by the Space Shuttle, and would be serviced on a regular basis by the Shuttle crews.

But the Leasecraft project, well-intentioned and full of promise, failed. Many of the factors that inhibited this venture are as relevant today as they were 20 years ago.

Background

The early 1980s were heady times for space commercialization. The success of the Communications Satellite Corporation (COMSAT) had demonstrated the viability of the commercial satellite businesses in space. The Space Shuttle Columbia lifted off in April 1981, with a second Shuttle launch in November of that year. In 1982, there were three more flights to deploy satellites. These launches marked the end of flight-testing and the beginning of a new era known as the Space Shuttle Age. This was the time when the pioneering ideas that emerged in the late 1950s from Lockhead Corp's Missiles and Space Division (Kramer and Byers, 1958) of a space station began to be taken seriously.

In December 1982, Fairchild Industries, Inc. announced that it was dividing its Space and Electronics Company into two entities, an initiative designed to capitalize on the anticipated growth in the commercialization of space. Fairchild Communications and Electronics Company would retain responsibility for existing military contracts for aircraft data communications and commercial satellite Earth station construction and components.

Space activities would become the responsibility of Fairchild Space Company. John Townsend, president of Fairchild Space Co., and a former Director at Goddard Space Flight Center, announced that the company would concentrate on developing an orbiting platform called Leasecraft (Figure 4.1). Leasecraft would be available to private commercial ventures for affordable materials processing and Earth observation, and would be launched and tended by the astronaut crew of the Space Shuttle.

A couple of months earlier, in October, 1982, Fairchild Industries had signed a morandum of understanding with NASA for launch and service support of the first of a fleet of 10 Leasecraft space platforms (*Defense Daily*, 1982). At this time, Fairchild estimated that it would spend $100 million to develop the first Leasecraft, and hire up to 200 new engineers, scientists and technicians for its Germantown plant. NASA would launch the first Leasecraft and service it about six months later in exchange for payload space, providing Fairchild with significant savings in launch costs (Lowndes, 1982).

The Leasecraft satellite platform would be able to support a production facility that would manufacture pure pharmaceuticals, perfect crystals for semiconductors, truly round ball bearings, and other metals and alloys. The initial rental estimate was $5 million a month (later reduced to $4 million a month). Fairchild expected to operate each of the crew-tended orbiting platforms for 10 years, with the first launch in 1987 and nine more launches within five years. The space factories would pay for themselves in three or four years, assuring a profit of about $400 million for each platform over its life cycle, or a total of $4 billion for the entire fleet.

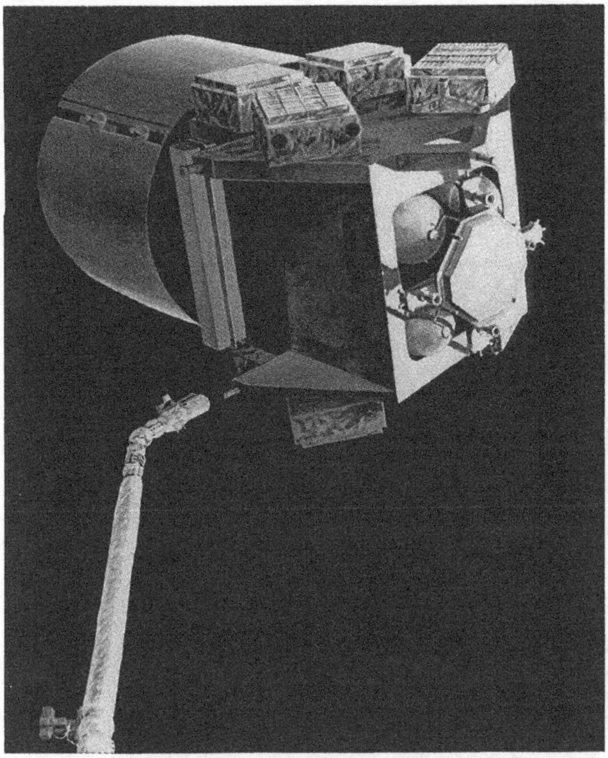

(*Source*: NASA History Office)

Figure 4.1. Proposed Leasecraft Satellite

After signing the Leasecraft agreement with NASA, Fairchild Space Co. hired former NASA Associate Administrator, John E. Naugle, as senior director of the Leasecraft project. Fairchild also tried to finalize an agreement with McDonnell Douglas and Johnson & Johnson, partners in a McDonnell Douglas developed process called 'free flow electrophoresis' which would use the low gravity of space to purify biological substances such as hormones and enzymes. The partners were also considering the possibility of manufacturing three dozen other possible pharmaceuticals, including an enormously profitable drug for anemia, a treatment for diabetes, and a possible treatment for AIDS, in a space factory. Metal processing and manufacturing computer chips were also thought to be good candidates for space factories (see Figure 4.2).

The 1982 NASA/Fairchild MOU indicated that first launch would occur in 1986 with estimated development costs of Leasecraft of $100 million. In September 1983, NASA Administrator James Beggs signed a joint endeavor agreement with Fairchild for Leasecraft (Knight, 1983). By then the schedule had slipped by a year and the projected development cost had doubled to $200 million. Nevertheless, in early 1984, Fairchild Space Co. was ready to guarantee a fixed price rental of $4 million a month for Leasecraft. NASA would launch one platform on the Space Shuttle and re-supply it after six months in exchange for a portion of the space factory for NASA experimentation, presumably in preparation for the anticipated Space Station.

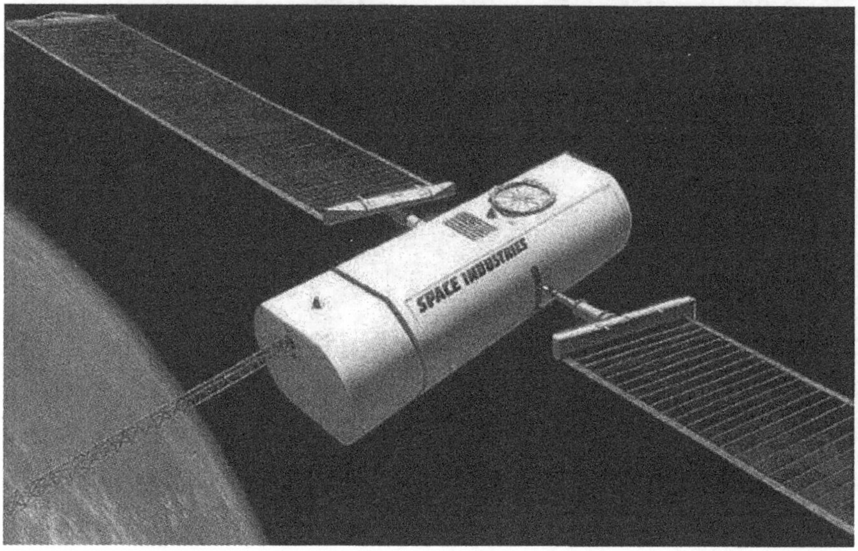

(*Source*: NASA History Office)

Figure 4.2. An Industrial Space Facility

Politically, the time appeared to be ripe. In his 1984 State of the Union address, President Reagan stressed the role of private enterprise in the commercial exploitation of space and announced approval for the development of a Space Station, an orbiting laboratory that could do all that Leasecraft promised, and much more. The Space Station would be another instrument of the Cold War, in competition with the Soviet Mir space station. If the US could manufacture drugs and computer chips on a large scale in space, the nation's technological prowess, fueled by private industry and supported by the government, would be unmatched.

NASA's agreement to launch and service Leasecraft was viewed as a prime example of government-industry cooperation. However, some questioned the federal space agency's ability to deliver on its promise to launch and service the first Leasecraft, much less extend it to a second or tenth launch and accomplish dozens of servicing missions. The Space Shuttle manifest was filling up with payloads from commercial satellite customers. The Hubble Space Telescope was nearly ready to secure a launch date, as were many space science experiments. Even though the Shuttle program was in its infancy, the demand for manifest slots was a harbinger of things to come.

Despite broad-based interest, favorable political alignment, and eagerness to pioneer biological and materials processing in space, Fairchild failed to induce McDonnell Douglas, Johnson & Johnson, or anyone else to sign up for Leasecraft. It was a spacecraft without a mission. William Fulwider, a spokesman for Fairchild Industries, noted that although there was a lot of interest in Leasecraft no one seemed to want to be first to climb onboard. The August 1983 announcement of Max Faget, former director of engineering at NASA's Johnson Space Center, that his company, Space Industries, Inc. planned to build an industrial space platform for materials processing and manufacturing had caused some potential customers of Leasecraft to hesitate. Shuttle crews would harvest, resupply and service the Space Industries, Inc facility until the Space Station crew could take over, just like the Leasecraft.

By the end of January 1984, NASA had also signed a memorandum of understanding and a joint endeavor agreement with Space Industries, Inc. for its plan to build the world's first crew-tended space platform. Robert Freitag, Deputy Director of the Space Station Task Force, said this was consistent with both NASA's and Space Industries Inc.'s mutual interest but must have been confusing to those who may have contemplated a lease with Fairchild. (A *Memorandum of Understanding* was signed by John Hodge on August 20, 1982.) The January 1984 announcement of the approval of the Space Station may have also made some potential customers reluctant to commit to the Leasecraft platform.

More uncertainty for Leasecraft investors and potential customers came in June 1984 (Tucker, 1984). At a hearing on space commercialization on Capitol Hill, Representative Ed Zschau argued that subsidized launch services from NASA could destroy the private launch companies seeking financing to start up new companies. NASA Administrator James Beggs even said NASA would be ready to sell the Space Shuttle business to private industry by the end of the decade. He noted NASA already had some 'private sector interest' in such a privatization scheme. Rep. Zschau argued that NASA should focus on research instead of business, but Grumman Aerospace Corporation, McDonnell Douglas and Fairchild disagreed. Grumman said 13 major corporations were interested in Leasecraft experiments, but John Townsend warned that high Shuttle prices would dim this interest. It appeared that private launch interests were pitted against the space manufacturing/processing interests in the commercial arena.

In October 1984, the US Internal Revenue Service ruled that most tax incentives for research and development of American-made products within US borders did not apply to outer space. For Fairchild, that meant most of the $80 million in Leasecraft research and development could not be deducted as a business expense. Fairchild would not be eligible for the 10% investment tax credit on new equipment, and the newly enacted five-year depreciation on new equipment would not apply to the orbiting factory. The free ride to outer space and the *quid pro quo* servicing done by the Shuttle crew would also be treated as taxable 'benefits.' In addition, goods manufactured in space would be treated as 'imports' under US trade law (Knight, 1984).

'Times are tough at Fairchild,' one analyst said in late May 1985. Nevertheless, the company chose to gamble on its aerospace and high technology operations by selling its consumer products division. On May 28, 1985, Fairchild announced it would sell VSI Fasteners Co. and VSI Hardware Co. (Schrage, 1985). The VSI group had accounted for half of Fairchild's pretax profit the year before, but the company needed cash for the satellite and Leasecraft businesses.

In August 1985, NASA Administrator Beggs and Space Industries, Inc.'s Max Faget agreed to have the Space Shuttle carry the Space Industries, Inc. spacecraft, now called the Industrial Space Facility, into Earth orbit. This facility, estimated to cost somewhere between a quarter and a half-billion dollars, would house a chemical factory.

Leasecraft's challenges continued to multiply. Potential customers had learned that they could not get insurance coverage on Leasecraft experiments (Covault, 1985). Several launch failures had caused insurance rates to escalate to a point where as much as 20% of the launch cost was for insurance. Without a customer for its proposed space factory – in addition to other difficulties – Fairchild ceased further Leasecraft development in the fall of 1985.

After failing to obtain insurance from private sources, Fairchild had asked NASA to act as insurer by underwriting Fairchild's loss in the event of a shuttle launch failure. Fairchild also wanted NASA to agree to $75million of 'termination liability insurance' in case the Agency 'decided to withdraw from the leased platform services it would be paying for beyond the joint endeavor provisions.' NASA did not agree to these insurance provisions. NASA responded with an offer to reactivate the joint endeavor agreement with Fairchild if Leasecraft could find customers and insurance coverage.

In January 1986, the Space Shuttle *Challenger* exploded a minute into launch. For all intents and purposes, the Leasecraft program was dead. Max Faget of Space Industries, Inc. hoped the *Challenger* accident would just stretch out the development cycle of the Industrial Space Facility rather than cause the project to be terminated, but that did not happen either, despite a joint venture with Westinghouse in September 1986 (Isikoff, 1986). The hope for an orbital manufacturing facility was lost.

Limiting Factors

As with most failures, that of the Leastcraft project was not the outcome of one thing but rather the result of a multiplicity of 'limiting factors.' Some of these – including time, transportation, competition, technology, materials processing, financing, insurance, government funding for the Shuttle, power and even floor space – are discussed here.

Transportation

Transportation is the main constraint that limits the development of commercial space products. The single transportation vehicle for materials to be returned to Earth was, and still is, the Space Shuttle. No other viable method (other than the 'parachute and capture' system used by the Air Force to recover film) is available. There is no indication that any commercial venture has ever attempted to use the Air Force recovery system.

The problem was summarized by John Naugle, director of the Leasecraft project, who believed that an 'assured source of transportation' is essential to the viability of any commercial production operation in space. Naugle observed:

> ...(a commercial customer) trying to make a profit in space is dependent upon the Shuttle. The shuttle is not a commercial transportation system. A commercial customer has...a priority behind NASA and DOD [Department of Defence] payloads. In addition the fate and schedule of the Shuttle is not determined by the marketplace but by the political process...Commercial communications and earth observation systems work because their product is transported back by radio waves – they only need one way transportation.

Shuttle Competition

Charlie Walker, who flew on three Shuttle missions as a payload specialist, commented:

> At that time [1979-1981], there was no perceived competition for Shuttle resources to fly industrial research and development. NASA's only 'commercialization' (i.e., 'industrialization') support for space processing was in the Electrophoresis Operations in Space (EOS) project. If one assumes that 'competition' should be viewed as payloads needing manifest slots, then all Shuttle payloads (in the crew compartment or laboratory) requiring crew attention could be considered competition.

Materials Processing

The number of viable, nearly ready, marketable products that could be processed in space is very limited, with only a few thought to be actually feasible. One of these was the McDonnell Douglas electrophoresis operation in the Electrophoresis Operations in Space (EOS) program – the prime customer in Fairchild's business plan. In addition to the EOS program two other initiatives seemed to be likely customers for an orbiting space manufacturing facility – the 3M Corporation's 'polymers in space' project and a semiconductor processing initiative being considered by Grumman Aircraft Corporation. Although several other companies (one source said thirteen) were interested in the potential of space manufacturing processes, they were not willing to commit funds for their potential products.

Time

The Soviet Mir was the only space station in the 1980s. The Mir space station was small, and its time on-orbit was very limited, especially for conducting R&D experiments for the development of commercial products. With a Shuttle flight limited to about two weeks on-orbit, competition for Shuttle crew time further constrained potential commercial R&D experiments.

This problem should be mitigated by the capabilities of the International Space Station to conduct R&D experiments in the future, but competition for on-orbit time will still be fierce for US commercial ventures. US science and technology concerns will have to compete continually with the International Space Station partners who have contributed to the Space Station's design, launch and assembly.

Financing

Private financial investment in both the development and operations of space-based ventures has always been difficult to obtain because of the long time span to reach profitability. Although Fairchild stated early in the program that Leasecraft would produce a profit as early as three to four years after the first spacecraft was launched, others estimated that it would take ten years for Leasecraft to become profitable.

Insurance

Commercial space development is inherently financially risky. The Fairchild company's ability to obtain catastrophic insurance cover for the Leasecraft program was one of the main reasons for the failure of the project (Covault, 1985). It was also unable to reach an agreement with NASA on the space agency underwriting either the entire $100 million of catastrophic loss insurance or the $75 million termination charge. At the time, several Ariane launch vehicle failures had made the insurance companies extremely risk averse, and the price of insurance, if obtainable, was very high.

Government Funding

Funding for the development of the Shuttle was restricted and required design tradeoffs that increased recurring operations costs. Limited development funding for launch vehicles and the enormous expenses of the operations infrastructure have severely constrained potential commercial space ventures to this day. This is particularly true in cases where a return payload is required and crew time is necessary for product development and manufacturing activities.

Essentially, the US Congress was not interested in developing a space transportation infrastructure in support of commercial space that would be comparable to the development of the interstate highway system.

Technology

Leasecraft servicing was dependent on extravehicular activity (EVA) by astronauts. In the 1980s, little had been done to adapt robotics technology to permit products to be harvested from Leasecraft. At the same time, EVA training is costly and dangerous for Shuttle crew members. Furthermore, since Leasecraft was to use existing 1970s technology, its capabilities would have been more than 10 years old at the time of first flight, potentially restricting its productivity and commercial viability. Improved robotics would have reduced the number and difficulty of EVAs and made it much easier to transfer materials between the Leasecraft and the Shuttle.

Power

A major limiting factor in space experimentation is the availability of adequate energy to power furnaces for materials processing and to do basic laboratory work. In the early 1980s, concerns about the safety of nuclear power, essentially eliminated it as a viable energy source for research in space. However, against this, nuclear energy may be required to provide the power needed to drive space experimentation programs for commercially viable products and new discoveries in the future.

Floor Space

On-orbit volume available to conduct experiments is also limited. The International Space Station has much more space for experimentation than the Shuttle, but Space Station volume is still constrained. With almost 20 nations conducting proprietary experiments in a limited volume of space, hoping to discover innovations that will recover at least some of their investment, floor space on the Space Station is very expensive.

Failure Analysis

In addition to the limiting factors discussed above, there are a number of other reasons why the Leasecraft Project might fail even today. Almost any one of these factors could have ended the venture, and the combination of multiple problems was fatal.

Lack of Customers

The 'free flow electrophoresis' process sponsored by McDonnell Douglas did not develop as a viable commercial product to be processed in space. Ground-based experiments achieved results almost as good as the space-based production expected on Leasecraft, but at a fraction of the cost. As a result, Fairchild's efforts to commercialize a space-based process were frustrated by the development of an acceptable Earth-based process. But even if the process had remained viable for space production, it is possible that Leasecraft would not have been financially viable.

There are a number of important prior conditions necessary to make a venture like Leasecraft financially viable. Opinions differ, for example, on whether customers exist for a resource similar to Leasecraft in the future. Even today many people involved in the Leasecraft Project doubt that commercial ventures would be feasible.

Jim Rose, formerly with McDonnell Douglas, said:

> The same problem exists today. There are no ready users for the Leasecraft support. In 1985 McDonnell Douglas was thinking on-orbit production by 1988. As of now, there are some promising technologies that have reached the development stage with nowhere to go to complete the effort to bring them to the stage they need to be at to fly on a Leasecraft and Industrial Space Facility or any other free flying platform.

John Naugle, former Leasecraft Manager at Fairchild, said:

> I do not think that Leasecraft would work today. You need a commercial product with an annual revenue on the order of $1 billion that absolutely cannot be manufactured on the ground for the foreseeable (at least 10 years) future. It is extremely unlikely that some clever scientists and engineers couldn't figure out some way to manufacture such a product on the ground in 10 years. If you don't have such a commercial product, then you need a guaranteed government contract to cover the costs of development.

The development and implementation of space-based commercial products still have a long way to go. A NASA (1987) report in the late 1980s concluded '…that with a small number of experiments, the ISF [Industrial Space Facility] did not offer unique capabilities,' whilst Pike (1988) argues that 'Clearly, the nation's need for the commercially developed space facility is far from proven.'

Fairchild's Financial Condition

The precarious financial position of Fairchild during the mid-1980s was another possible reason for the termination of the Leasecraft Project. Future transportation commitments using the Shuttle were limited and uncertain, and Fairchild was unable to reach an agreement with NASA on insurance and termination liability. Fairchild was facing an unstable financial future, having also just lost $300 million on a commuter plane project. Faced with all these pressures, Fairchild was forced to put the program on hold. Wolfgang Demisch, noted aerospace analyst, said that Leasecraft was in trouble because Fairchild was in trouble, even before the *Challenger* accident.

Shuttle Demand

Before the *Challenger* accident NASA's restrictive Shuttle policy for payloads in the 1980s made flying on other vehicles difficult. This increased the demand on the shuttle.

To sustain 10 Leasecraft in orbit and service each one every six months, almost two Shuttle flights per month would be required. The joint endeavor agreement committed NASA to supporting only the first two flights: one to deploy the Leasecraft from the Shuttle and another to service the Leasecraft and its payloads after a six-month testing and demonstration period. Supporting 10 Leasecraft would have strained the Shuttle manifest.

In addition to the science and technology payloads vying for slots on the Shuttle, the Industrial Space Facility was still in development, and the Space Station had just been announced. Both of these endeavors would have significant commercial components, and would place heavy demands on Shuttle services and operations. If all of these projects developed, the Shuttle manifest would be seriously oversubscribed.

Lack of an Adequate Orbital Return System

Unrealistic expectations also may have contributed to the downfall of Leasecraft. In 1982, Congress and the White House were strongly supporting proposals to exploit the use of space for commercial activities. The communications satellite business was a big success, and it was expected that other commercial activities would also be successful.

Unfortunately, no one seemed to have thought much about getting products back to Earth, or of the enormous cost of manned activities (especially for EVA) required to service the manufacturing facilities. In the 1970s and early 1980s, no one, including NASA, seriously questioned whether the Shuttle would ever actually be able to meet the expectations of one flight per week from the East and West coasts. The root cause of the failure of Leasecraft may have been Congress's failure to fund the development of a Shuttle that would really be reusable and inexpensive.

Cost

Fairchild estimated that payload processing on Leasecraft would cost $4 million a month, compared to a Shuttle cost in the order of $75 million to $100 million per flight. Jim Rose estimated that it would take a quarter of the capacity of the Shuttle to service the EOS on the Leasecraft. If the Shuttle had been able to fly once a week from either coast, would $24 million for a six-month manufacturing period have been adequate to cover the expenses of the Shuttle alone? In retrospect, it seems optimistic, and it is doubtful that this amount would also have covered the amortization cost of development.

The development cost for the Leasecraft was originally estimated at $100 million, but was revised to $200 million when the joint endeavor agreement was signed a few months later. Operational costs were not estimated, even though operations are very expensive for manned flights and space operations in general. The use of EVA required extensive and expensive training in order to move manufactured products from the Leasecraft on-orbit to the Shuttle, and to supply new material for processing on the Leasecraft platform. According to Jim Rose, McDonnell Douglas never obtained a cost figure for crew time and training. Even though the advertised cost seemed low, to a prospective user, the manufactured product still had to have a very high price per pound to justify the expense.

The tax situation added another expense to the process of space manufacturing. The IRS ruled that US companies could only claim tax deductions and credits for investments whose 'physical location' was within the nation's borders.

In summary, it appears that a $4 million per month price to the customer for amortization of the development cost, the Shuttle flight cost and the operations cost was optimistic.

The Challenger Accident

The *Challenger* accident in January 1986 was the final blow to commercial plans to utilize space using the Shuttle. Fairchild had suspended Leasecraft development in November 1985. The McDonnell EOS proposal had been withdrawn because the desired product could be produced on the ground. In addition, McDonnell Douglas had also been negotiating with Space Industries at the same time to secure a place on its Industrial Space Facility, which also would have been serviced by the Shuttle. Fairchild had not been able to secure a firm customer.

Did the unfounded optimism of the early Shuttle program guarantee that programs like Leasecraft would fail because of the inability to service the facility at a rate necessary to support a commercial venture? Some believe that the failure could be directly traced to the unreliability of the Government sponsored space transportation system – the Shuttle. However, at the time the Leasecraft Project was being formulated, the Shuttle had just started flying, and nobody had realistic expectations for its potential.

Conclusions

Analysis of the Leasecraft program supports the conclusion that the Leasecraft Project would have failed even without McDonnell Douglas withdrawing from the program. The problems Fairchild experienced with the Leasecraft program still exist and would impede almost any new commercialization venture.

NASA entered agreements with both Fairchild and SSI before verifying that the resources available for commercialization of space would be adequate. The Shuttle, which had the only orbital return capability for made-in-space processed products, was so oversubscribed that it is doubtful it could have provided enough payload slots for Leasecraft. Before any company undertakes a commercialization venture today, NASA should demonstrate that it has a well-integrated payload planning policy and implementation plan in place. It is also clear that the use of EVA resources was much too extensive.

Too much emphasis was placed on cheap access to space. Any transportation system in its early development is going to be expensive. The cost of delivery to space should not be compared to ground transportation, for which a well-planned infrastructure has been built over time. Until there is a more robust space transportation system the government must subsidize launch and servicing costs to some extent in order to establish a viable commercial space sector.

The competition between science and technology for space project funds has affected commercialization of space because of the relatively small amount of funding awarded to technology projects that support commercialization (Washburn, 1984). The demands of science have limited the availability of Shuttle manifest slots for commercial ventures. This has been a core argument since the early days of the aeronautical industry. In the 1920s, NASA's predecessor, the National Advisory Committee on Aeronautics (NACA), argued successfully for the development of technology. NACA gave priority to engineering over science throughout its existence (1913-1958). This is not true of NASA today.

The Shuttle, and access to the Space Station, are essential to the development of commercial processes in space. The *Challenger* accident delayed the start of commercialization by limiting the use of the Shuttle for private industry. NASA policy and the scientific community's desire for more control of the payloads have slowed the pace of commercial uses of space. The Shuttle was not allowed to fly commercial payloads after the *Challenger* accident.

The technology used in the Leasecraft was current but not cutting edge. The Leasecraft program didn't use robotics, resulting in excessive dependence on EVA. Fairchild was proposing to build the Leasecraft using existing technology that had been pioneered for the earlier Multi-Mission Spacecraft (MMS) system.

NASA policy seemed responsive and reasonably clear. However, there was, and continues to be, an uncertainty in NASA funding levels that makes companies wary of entering into a venture that depends heavily on government resources. Commercial space companies need secure on-orbit facilities to further develop technology that can be commercially viable.

Acknowledgments

The author would like to acknowledge the help given by NASA Headquarters History Office. Charles Walker, Jim Rose, John Naugle and George Mason University faculty and staff.

References

Covault, C. (1985) 'Lack of Insurance, Customers Halts Fiarchild Leasecraft', *Aviation Week and Space Technology*, November 11, pp. 16-17.

Defense Daily (1982) 'NASA Would Launch Leasecraft in Exchange for use', October 12, p. 190.

Isikoff, M. (1986) 'US Firms Unveil Plan for Space Lab', *Washington Post*, September 30, p. C3

Knight, J. (1983) 'Factories in Space', *Washington Post*, September 23, pp. C1/C3.

Knight, J. (1984) 'Taxes Ground Space Pioneers', *Washington Post*, October 15.

Kramer, S.B. and Byers, R.A. (1958) *Assembly of a Multi-manned Satellite*, LMSD Report No. 48347, Washington DC.

Lowndes, J.C. (1982) 'Fairchild, NASA Agree on Leasecraft', *Aviation Week and Space Technology*, October 18, pp. 14-15.

National Aeronautic and Space Administration (1987) *Report on the Potential NASA Utilization of the Space Industries Partnership Industry Space Facility (ISF)*, NASA, Washington, DC.

Pike, J. (1988) 'Let's Find out who will use it Before we Waste a Billion Dollars', *The Scientist*, vol. 2(12), p.13.

Schrage, M. (1985) 'Fairchild to Sell Consumer Units; Will Unload Hardware, Fastener Divisions for $35Million', *Washington Post*, May 29, p. F1.

Space World (1984) 'Developing and Marketing Start on First Satellite for Hire', January, pp.10-11.

Tucker, E. (1984) 'Space Shuttle Pricing Debate', *Washington Post*, June 20, p. F1.

Washburn, M. (1984) 'What's A Space Station Good For? We Should Be Aiming Higher – Toward Space Tugs, Asteroids And the Planets', *Washington Post*, April 1, p.C1.

Chapter 5

NASA and the Evolution of the Hush Kit: A Technical Solution to a Social Problem

Lawrence S. Jessie, Francis T. Hoban and William M. Lawbaugh

Introduction

Since the beginning of aviation, aircraft noise has been a problem. Noise pollution is not simply an environmental problem, but may also be a psychological problem, a medical problem and a sociological problem. It can also affect productivity and result in a heavy financial burden on society (Gillen, 2003). Fundamentally, though, aircraft engine noise is a technological problem and NASA's approach to aircraft engine noise abatement was to find a technological solution.

Modern aircraft engines are much quieter than they used to be. In fact, on take-off as well as landing, more noise is produced by the swoosh of wind against the plane than by the noise of the engine, even in reverse. But what about older airplanes – the planes that created the problem of noise pollution in the first place?

Newer aircraft have built-in noise abatement design, but older aircraft have to be retrofitted with quieting devices known as 'Hush Kits', new engine housings (nacelles) lined with sound absorbing material. This study examines the noise pollution problem created by aircraft engines and describes the medical and social repercussions. It discusses political attempts to deal with an angry airport neighborhood populace, and the legislative and administrative initiatives that address the problem. Finally it describes the evolution of NASA's technological solution to one aspect of aircraft noise pollution.

Aircraft noise reduction has been a contentious issue since the first days of commercial airport construction; there have been lawsuits against aircraft noise since the 1920s.

In the 1960s the aeronautical industry boomed, and the nuisance of aircraft noise intensified. People living near airports not only began to complain vociferously about increasing noise, but also began to demand that government and industry do something about dangerously high levels of noise pollution.

NASA took up the challenge of reducing jet noise in 1971, having already spent several years researching the technology needed to build quieter engines and quieter take-off and landing procedures. NASA administrator, James Fletcher, proclaimed: 'It is no longer enough to think in terms of more power, more lift, more speed. If we wish the taxpayers to continue to support civil aviation, we should take as our own motto, "Fly Quiet"' (*NASA News Release*, 1971).

Two years earlier, NASA engineers at the Langley Research Center in Virginia had announced better than expected results in a retrofit of the noisy Pratt and Whitney engines that powered the four-jet DC-8 and the popular Boeing 707, a retrofit designed to reduce the noise generated by these aircraft. The cost of the retrofit also came in lower than expected, just $1 million per jet aircraft. The devices were acoustically treated aircraft engine nacelles, later known as Hush Kits.

Although the news of the success of the Hush Kits spread rapidly, acceptance and implementation by the Nixon Administration came more slowly. In 1968, the last year of his administration, President Johnson had signed into law a bill ordering the federal government to develop stricter policies on airport noise abatement. This legislation initiated federal efforts to find solutions for the problem of aircraft noise. However, within three years the economy had soured, and the new administration and Congress were more interested in ending the war in Vietnam than reducing jet engine noise.

But in 1971 NASA looked at the challenge of airport noise abatement as an opportunity, not just a problem. At a NASA-sponsored three-day conference on technology for civil aviation, James Fletcher told industry executives and FAA officials that 'the environmental issue of noise "provides a showcase for technology" at Langley Research Center, especially the NASA nacelle retrofit test. "Development of Hush Kit technology"', the NASA Administrator added, provides 'an opportunity to apply technology directly to the solution of an environmental problem of increasing public concern' (*NASA News Release*, 1971).

NASA took up the challenge, partnered with both Boeing and McDonnell Douglas, and met the Johnson Administration mandate for quieter aircraft. The evolution of this effort serves as a case study of government and industry working together to develop a technical solution to a social problem.

Similar partnerships offer much potential for other efforts to solve the environmental, health, psychological and sociological problems that confront society. By working with two competent independent contractors on the aircraft noise problem, NASA was able to launch an entirely new industry, and provide a model for subsequent endeavors in the privatization movements of the 21st century (Hoban et al, 1997).

Public Purpose and Social Responsibility

Aviation noise abatement efforts date back to the 1920s when both commercial travel and cargo transport were literally taking off. As the public began to fly routinely, the air transportation industry began growing exponentially. In 1926, Congress passed the Air Commerce Act, the cornerstone of the federal government's subsequent efforts to regulate civil aviation. The Act tasked the Secretary of Commerce with fostering air commerce, certifying new aircraft, licensing new pilots, establishing airways, and creating, issuing and enforcing air traffic rules, including noise regulations. William MacCracken, chief lobbyist for creation of the new agency, was the first head of the Aeronautics Branch in the Department of Commerce.

Initial complaints came from those who considered airplanes to be a hazard or an annoyance. Some had legitimate fears that over-flights would damage their homes, barns or cattle. The unaccustomed sound of airplane engines whirring and sputtering overhead was annoying to those who lived near an airfield.

For the first time, two major questions arose for public debate. Do noisy airports serve a public purpose, or are they built and regulated with public funds primarily for the convenience of a few companies, stunt flyers and recreation? Second, who is responsible for aircraft engine noise: the fledgling aircraft industry, commercial interests or the federal government? It was the responsibility of the Department of Commerce's new Aeronautics Branch to respond to the emerging political and economic issues raised by decisions related to airport location, funding and expansion. Aircraft and airport noise were already becoming an important issue.

In the 1930s, during the economic turndown of the Great Depression, air commerce was viewed as an essential component of recovery. As early as 1933, federal funding of airport development was intended to help create jobs for American workers as well as to build military air capability. Two federal recovery programs, the Development of Military Landing Areas program and the Development of Civil Landing Areas program, led to the building of an incredible 584 airports. Both of these programs are still operating. From 1933 to the end of World War II, the federal government spent $912 million for airport expansion.

With the movement of population from the cities to suburbia, airport noise pollution became an even greater concern. In 1934, the Department of Commerce renamed the Aeronautics Branch to reflect its growing stature in the agency. The Bureau of Air Commerce encouraged airlines to develop a coherent system for air traffic control. In 1936 the Bureau took over the system and federalized the air traffic controllers. These pioneers used maps, blackboards and mental calculations to ensure the safe separation of aircraft flying in and out of the expanding airports and along designated air routes.

Fast growing cities such as New York, Los Angeles and Washington, DC, expanded their existing airports and built more. Towns began to build up around these and other airports as out-of-work mechanics found new employment in the commercial aircraft industry. And, as America began to prepare for its approaching involvement in a global war, the aircraft industry prospered like never before.

The aircraft industry boomed when America entered World War II after Pearl Harbor. Deafening bombers and troop cargo and transport planes roared overhead, built by Rosie the Riveter in the heartland of America. Airport noise pollution took a back seat to war preparations and conscription. At this time aircraft noise was a potent symbol of power and military might. More noise – jet engine noise – was on the horizon. In 1930, Frank Whittle of England patented the design for the jet propulsion aircraft engine. Whittle's design would become commercial and viable at the end of World War II.

After the war, noisy aircraft engines again became a problem to deal with, but other priorities took precedence. In 1946, President Truman signed into law the Federal Airport Act, which provided massive amounts of federal aid to fund the growth and development of the country's civil airports. Except for a brief hiatus in fiscal year 1954, the Federal Airport funding program continued for 24 years. Fueled mainly by Cold War fears and military readiness, aircraft technologies and airport infrastructure continued to expand, in spite of successive bouts of economic instability in the 1950s. The airline industry seemed to be recession-proof.

In 1957, the government created the National Aeronautics and Space Administration (NASA) in response to the Soviet launch of the first satellite, Sputnik, and the space race was on. For this decade, at least, the first 'A' in NASA seemed completely overwhelmed by the race to the moon. In 1966, Congress authorized the creation of a new Cabinet-level agency that would combine most other civil aviation responsibilities.

The US Department of Transportation (DOT) opened for business on April 1, 1967. On the same day, the Federal Aviation Administration (FAA), which had been established by the 1958 Federal Aviation Act, was officially named as one of several model organizations within DOT; one of its first tasks was to do something about the noise problems near airports.

In 1967 the only research on aircraft engine noise available to the FAA had been done by NASA, and its predecessor agency, NACA. Aeronautical acoustics work had been ongoing for several years at each of the three NASA Research Centers (Ames, in California, Langley in Virginia, and Lewis, now Glenn, in Ohio), including fundamental research on noise generation and propagation from aircraft components. The NACA/NASA acoustics research went back four decades. Initially this research concentrated on understanding the effects of noise on the human person in general.

Aircraft Engine Noise and the Human Response

Noise effects on humans are numerous and complex. There has been relatively little research done on noise or noise pollution, and the results are often inconclusive and certainly not universally valid. Different people perceive noise differently, due to individual emotional and physical variables. Although, it is difficult to determine an exact decibel level that actually harms human beings, most researchers believe that physical harm begins at 85-90 decibels. Some people can withstand much more noise than others. Some even seem to thrive on certain kinds of noise, often referred to as 'white noise', such as the sound of waves hitting the beach, music, or the sound of rain on a roof.

Using the NASA studies and those of others, the FAA studied the individual perception of noise, the variables affecting response, different sources/different human response, community hearing loss, and the application of these findings to aviation issues (Federal Aviation Administration, 1985; for more recent analysis see Schomer and Associates, 2001).

Perception of Noise

Individual perception of noise, the loudness of sounds, depends on several characteristics:

- Duration. The perception of loudness may depend on how long someone has to listen to a particular sound. Airplane noises can seem to get louder and louder over time.
- Frequency. To most people, bursts of sound between 2,000Hz and 8,000Hz seem louder than equivalent sounds heard at a steady level. The occasional roar of engines seems louder than a constant roar.
- Ascendancy: Sounds that rise gradually seem to be louder than those that are decreasing in level and this is sometimes actually used to societal advantage – i.e., sirens on a police cruiser or ambulance. People usually object more to the noise from aircraft takeoffs than from landings.
- Rate. A rapid increase in the sound pressure level seems louder than occasional, intermittent sounds. Impulsive sounds that peak rapidly, like a jackhammer or pile driver, are perceived to be louder than an equally loud crack of a baseball bat. The sudden thrust of an aircraft engine in reverse on touchdown can be perceived as jolting, arresting.
- Intensity: Listeners often perceive a slight increase in sound as a doubling of sound. A slight increase in aircraft engine noise (such as that made by an older airplane) may be interpreted as a massive increase. People tend to exaggerate intensity.

Physical Variables

Individual reaction to a specific noise depends on a variety of physical and emotional factors that may be unique to the sound situation or to the individual listener.

- Seasonal. Generally people object to loud noises more in the summer than the winter. This is probably due to more direct exposure to the noise – people are outdoors more in the summertime, and windows are open more often.
- Time. Noise is more stressful to people at night than in the morning or during the daytime. Some airports reduce flights during the early evening and ban them after 10 pm.
- Geography. People expect rural areas to be quieter than cities. As a result, people react differently to the same level of noise in different locations. A given level of noise is perceived as more objectionable in rural areas than in industrial areas.
- Duration. Most people think it is possible to adjust to an annoying source of noise. However, the evidence indicates that resentment of noise intrusions increases over time.

- Predictability. Unexpected noises, such as a sonic boom, are more upsetting than predictable noises.

Emotional Variables

Studies of individuals who live in communities near airports indicate that emotional variables are far more complex than physical variables in their reaction to airport noise. These include:

- Importance. People are more tolerant of neighborhood noise if they believe that the activity that is the source of the noise is valuable or important. Few people resented the noise pollution of industry and aviation during World War II.
- Relaxation. A noise disruption is more acceptable to people when working, playing or communicating than when trying to relax, rest or sleep. Most people resent aircraft noise more during early evening and night than during the daytime.
- Expectations. When an airport extension or flight pattern intrudes upon an area that had not previously been affected by aircraft noise, the emotional cost is greater than if the change were made to a neighborhood which has been subject to aircraft noise for many years.
- Fear. The individual who thinks he or she will be harmed by airport noise will be the first to complain. The actual harm may be real or imagined, but the fear itself is psychologically harmful.
- Sensitivity. People vary widely in their emotional reactions to noise intrusions, their physiological predisposition to noise, their memories of similar noises, and even their keenness of hearing.
- Helplessness. People who believe that their objections to noise are not being seriously considered by the authorities are likely to object more strenuously than those who feel their concerns are being heard.

Different Sources/Different Responses

A very wide range of activities in modern society (Table 5.1) generates noise that can be a nuisance. Aviation noise, along with the noise of urban traffic, is the source of more complaints than other types of noise. Commercial and industrial noise is somewhat more acceptable, although the reasons for this are unclear. Studies in scientific journals confirm that there is a difference in perception, but that it is statistically insignificant. There is also evidence of variations in the response of different nationalities to noise sources (can den Bergh *et al*, 1997).

Community Hearing Loss

A study sponsored by the Federal Aviation Administration (1972) found no difference in the physical hearing acuity of a group of people living near Los

Angeles International Airport and residents of a comparable community far away from the airport.

Table 5.1. Noise Levels from Different Sources (dBA Values Encountered in Daily Life and Industry)

Source	Decibels
Window air condition	55
Busy restaurant	65
Vacuum cleaner (at 10 ft)	69
Loud music in large room	82
Threshold (beginning of hearing damage if prolonged exposure)	85
Printing press plant	86
Heavy city traffic	92
Diesel engine (at 25 ft)	92
Busy saw	97
Power mower	98
Jackhammer	107
Jet airliner (500 ft. overhead)	115
Pain threshold	120

(*Source:* Federal Aviation Administration, 1985)

In 1974, a laboratory experiment was conducted near Los Angles International Airport to see if aircraft flyover noise would cause a temporary threshold shift, considered to be a precursor to permanent hearing loss. For six consecutive hours, two groups of young men were subjected to recorded flyovers of a maximum of 111 A-weighted decibels each. The flyovers were repeated every 90 seconds for one group, every 180 seconds for the second group. The measured temporary threshold shifts for each of the subjects were negligible, suggesting no danger of permanent hearing loss from high levels of aircraft noise. This study was replicated in Japan, with the same results. These studies have been cited in lawsuits involving aircraft engine noise.

Sleep Interference

Human beings can adapt to noise. Most people are able to sleep through normal household noises such as the heavy gong of a grandfather clock throughout the night. Some types of noise, even traffic noise, can lull some people to sleep.

Much research on sleep interference has been done at NASA. In 1983, the FAA asked acoustical scientists and engineers at Langley Research Center in Virginia to review the literature of sleep interference and to re-evaluate the weightings for nighttime noise events (Federal Aviation Administration, 1985).

Sleep is commonly divided into two distinct phases: REM (rapid eye movement) and the deeper NREM (non-REM). REM sleep reoccurs approximately every hour and a half in a healthy human in cycles throughout the night. These undisturbed cycles are essential to the restoration of the human body and mind, to the consolidation of memory of information gained during wakefulness (dreams), and, in children, to the release of growth hormones. Although most people believe that undisturbed sleep is necessary for a healthy life, sleep interruptions do not appear to affect mental and psychomotor performance significantly, unless sleep deprivation is prolonged.

NREM sleep is divided into four sub-stages; stage 4 is deepest. The amount of time spent in stage 4 NREM sleep decreases progressively with age. Since older people experience more occurrences of sleep disturbance than do younger people, aircraft noise could be more of a problem for the elderly than younger adults or children.

Sleep researchers define arousal from sleep as occurring when, a) within one minute of a noise stimulus, the subject's EEG pattern changes to that of wakefulness, or b) the subject gives some sort of motor response indicating wakefulness. Research confirms that fewer arousals are found in deep sleep than in light sleep. Only very high exposure to aircraft noise could cause arousal from stages 3-4 of NREM sleep.

Because of the cyclical nature of the two sleep phases and the variances due to age, a normal person's susceptibility to arousal from sleep varies throughout the night. However, since more time is spent in the deeper stages of sleep during the first half of the night than during the second half, airport use restrictions are more important for early morning flights.

NASA researchers also confirmed that most people do adapt psychologically to new environmental noises. This adaptation involves learning when and how often noises are likely to occur, and learning how to adjust reactions or responses to prevent sleep arousal. Adaptation to noise during sleep is also a constant. In one study they examined, for example, the cessation of landings at Los Angeles International Airport from 11 pm. to 6 a.m. had no measurable effect on subjects' reports of sleep disturbances. These studies suggest that psychological annoyance from the effects of sleep interference from aircraft noise is probably more significant than direct physiological consequences.

Finally, the NASA Langley researchers determined that sleep arousal does depend on age and the stage of sleep, but noise level limits do cover a range of 35 to 70 dB. Hospitals recommend an interior noise level of 34 to 47 dB. For other sleeping environments, the maximum acceptable level of noise intrusion is set at 55 dB.

With the noise value of 115 dB for a jetliner 500 feet overhead, clearly the airports and the airline industry had serious noise issues to contend with, and government efforts to deal with the problem of aircraft noise pollution in the 1960s had failed.

Regulation

The Aircraft Noise Abatement Act of 1968 was the first legislation that specifically regulated sonic booms and other aircraft noise. The Act distributed this regulatory authority among three federal agencies: the FAA, the Environmental Protection Agency (EPA) and the US Department of Transportation. Although the EPA Administrator was ordered to establish an Office of Noise Abatement and Control, the FAA was given authority to control over noise and noise abatement at US airports. The US Secretary of Transportation was responsible for establishing 'reliable and uniform systems of measurement for noise and noise exposure', and could disburse grants to soundproof schools and hospitals in a non-complying area, as well as to soundproof or acquire residential properties.

In 1972 Congress passed the Noise Control Act that explicitly recognized for the first time that noise was a growing danger to public health and welfare. The Act permitted citizen suits against the FAA for airport noise pollution and established the Quiet Communities Program that offers state, local and regional grants to study noise abatement, buy monitoring equipment and collect data.

In 1990, the Airport Noise and Capacity Act mandated specific noise-levels for aircraft. This Act required that 85% of a carrier's subsonic Stage 3 (above 75,000lbs) transport fleet to comply with FAA-mandated noise levels by the end of the decade. Heavy fines are authorized for noncompliance.

Thus, the federal government sought legislative remedies for noise pollution, just as the US Congress successfully sought to clean up the nation's waterways (Clean Water Act) and the atmosphere (Clean Air Act). The legislation and federal regulation now required the aircraft industry to build quieter engines by the year 2000. But the airline industry was still flying older aircraft like the 707 and DC-8, that were noisy but still in use. In order not to mothball these aircraft, they needed to find some way to bring these aircraft into compliance with the new noise regulations. NASA had been working on the problems of aircraft noise for years, and, working with independent private contractors, eventually developed a technological solution to the noise problem of older aircraft.

A Unique Partnership

NASA's Langley Research Center has studied noise generation and propagation from aircraft components, the effects of noise on people and, later, sonic booms since the 1920s (Boswinkle, 1968). By the 1960s, due to growing pressures from environmental groups, citizen groups, and airport managers, President Johnson established an industry-government committee under the Office of Science and Technology (OST) to develop a national noise-alleviation program. The committee's most important recommendation was to ask NASA to expand its longstanding noise research program.

In 1966, funding support for noise research programs was under one million dollars. This budget line was raised to $18 million by 1968 in order to carry out the OST committee's mandate. Specifically, the committee asked NASA to accelerate the noise alleviation research along three lines:

- The use of steeper approach angles to take noise away from observers.
- Developing quieter aircraft engines.
- Applying sound-suppression techniques to engines and nacelles.

The first objective, improved flight procedures, was carried out by the FAA and NASA, using 'a two-segment approach procedure which would have an aircraft approach the airport at an altitude higher than is the practice today until it is approximately three nautical miles from the end of the runway at which point it would assume a normal approach glide slope angle'. Consultants Bolt, Beranck and Newman, Inc. (BB&N) reported that the 6% climb gradient takeoff procedure and 6°/3° bent glide slope approach procedure would reduce noise by 20% on landing for larger aircraft, and 25% on takeoff.

The second objective, called the Quiet Engine Project, was picked up by NASA's Lewis Research Center in Cleveland, Ohio. Lewis (now Glenn) combined all known noise control techniques into an engine with 20,000 lbs of thrust operating at high subsonic speeds. Their Quiet Engine would be 20 PNdB (Perceived Noise Decibels) quieter through better design of fan, compressor, burner, turbine and exhaust nozzles. Lewis issued key contracts to General Electric for this research, while Boeing studied compressor blade noise reduction.

Following early work by the Environmental Protection Agency (1973), BB&N in their analysis of the noise reduction programs, concluded that the results of the research on the third objective, sound suppression, could provide the most cost-effective solution for the near term – 'The nacelle retrofit program results in most of the achieved reduction', and the Quiet Engine costs of more than $4 million per aircraft were not 'economically reasonable'.

On August 30, 1966, an RFP called 'The Study and Development of Turbofan Nacelle Modifications to Minimize Fan-Compressor Noise Radiation', was issued by NASA to Boeing, Douglas, Lockheed and General Dynamics-Conair. The RFP called for a 15 PNdB reduction in flyover noise, no undue workload for crew, and economic viability. A month later, Boeing and McDonnell Douglas were selected for parallel development programs including nacelle redesign, ground run-up tests, analysis, wind tunnel tests, flight tests, performance calculations, and economic trade-off studies. All aircraft noise alleviation data generated prior to and through completion of the two parallel contracts would be made available to NASA to use and disclose at will in order to assure that public money was spent in the public interest, not proprietary corporate interests.

In less than two years, the research on nacelle acoustic treatment had yielded promising results. Ground test studies by McDonnell Douglas showed fan noise reductions of about 12 PNdB using sound absorptive liners and concentric inlet rings. The acoustic materials were made of fiber metal fabricated by Huyck Metals Co. Tests were also underway to document static strength, fatigue life, dynamic behaviors of various densities, fuel combustion, air oxidation, water saturation, salt corrosion, cleanability and freeze-thaw cycling (Office of Noise Abatement, 1968).

Boeing's research was also promising. Jet engine noise reductions of 10 to 12PNdB were measured with acoustic liners of non-metallic fibers. Boeing researchers were examining the effects of acoustic treatment on engine and aircraft weight, performance and operating costs. By early autumn 1968, Boeing and Douglas had turned over 35 aircraft noise research reports to NASA, each of which were available in abstract form in *NASA Scientific and Technical Aerospace Reports* (STAR) and microfiche form to entrepreneurs and investors.

While the NASA research was underway, the Johnson Administration worked to get Congress to act decisively on airport noise reduction. Finally, near the end of his final term, Johnson signed into law PL 90-411, the Aircraft Noise Abatement Act, on July 21, 1968.

Less than a half-year later, outgoing Acting FAA Administrator David Thomas announced proposed noise standards for new aircraft: a limit of 108 EPNdB (Effective Perceived Noise Decibels) measured at 3.5 nautical miles from takeoff and one mile from threshold of the runway for landings.

However, the lingering problem of noise reduction of existing aircraft became a hot political issue with the incoming Republican Administration. Kent (1980) describes the political situation at the time:

> While the initial noise rule proved to be uncontroversial, the question of retrofitting the current generation of jet transports did not. The issue came alive when the NASA nacelle test program, contracted out to Boeing and McDonnell Douglas, began producing results that were much better than originally expected. By the spring of 1969, it appeared that a 15-PNdB reduction in noise level might be achieved through a retrofit of the JT3D Pratt & Whitney engine with acoustically treated nacelles. These engines powered the B-707 and the DC-8 four-engine jets. Early cost estimates for the retrofit ranged up to $1 million per plane.

NASA held another conference at Langley Research Center to share the results of their research with contractors, scientists, engineers and other federal agencies. This research included the final reports on the McDonnell Douglas and Boeing investigations and a report on the Nacelle Acoustic Treatment Program done by Pratt and Whitney (a division of United Aircraft Corp.), dealing with 'noise attenuation in acoustically treated ducts'.

In addition, there were 44 papers (now made available from the federal Scientific and Technical Information clearinghouse) dealing with the aircraft noise problem from other contractors, made public for the first time. This unprecedented disclosure had been built into the original NASA contracts. As part of the contractual arrangement with McDonnell Douglas and Boeing, the two companies agreed to make available to NASA, for distribution to industry, the results of all their previous work pertaining to aircraft noise alleviation. Additional reports, then being reviewed by NASA, would also eventually be abstracted to NASA *STAR*, available at nominal cost. Thus research on reducing aircraft noise that had previously been proprietary was now available to all interested parties. By making the research available, NASA hoped to stimulate the Hush Kit industry and reduce the cost through information sharing and entrepreneurial competition. These contract provisions accomplished several benefits:

- Full disclosure would affect further private and government research on noise abatement.
- Duplication of effort was no longer necessary to test certain technologies.
- In R&D organizations, the research phase would be shorter and the development phase quicker.
- A new industry was launched for the benefit of all who live near airports.

The president of the Airport Operators Council International (AOCI) felt that future airport construction and expansion depended upon efforts to reduce aircraft noise. AOCI President Matthias Lukens wrote to FAA Administrator John Shaffer: 'If we do not solve these existing [noise] problems, we may never get a chance to show what we can do with new airports because the public will not permit them to be constructed... Therefore, the magic word is "retrofit." And it would be a bargain despite the expense' (Kent, 1980). To retrofit the existing fleet of older US aircraft would cost hundreds of millions of dollars. So, in spite of the promising NASA research on nacelles and the mounting public pressure to reduce airport noise, the FAA chose to focus on the quieter engines of the future rather than fix the noise of older turbofan engines.

The retrofit program languished for two more years. Airline industry opposition and the FAA's go-slow policy combined to halt the solutions found by NASA, despite public outcry about airport noise pollution. But Congress continued to pressure the FAA to implement the 'retrofit' solution proposed by NASA. In February 1970, just a few months after the final report on the 'NASA Acoustically Treated Nacelle Program' was issued, a group of concerned members of Congress met with FAA Administrator Shaffer to discuss 'technical problems on our retrofit rulemaking'. Shaffer indicated that the NASA acoustic development program 'had not provided information related to airworthiness requirements of the retrofit configurations nor did that program address the feasibility of retrofitting JT3D powered aircraft'.

On June 23, 1971, Congressman Joseph Addabbo, chairman of the powerful House Commerce Committee, once again questioned the FAA's 'regulatory action and rulemaking schedule for the reduction of airplane noise levels through "retrofit"'. Shaffer reminded Addabbo that the FAA's rulemaking schedule discussed by the Congressional delegation in February 1970 was contingent upon Congressional approval of $3.5 million, but that only $1 million was approved. He apologized for the 'schedule slippages' but hoped: 'the publication of an acoustic retrofit regulation would be possible by the end of 1972'.

The Administration continued to oppose the NASA solution. Civil Aeronautics Board Chairman Secor D. Browne addressed the AOCI annual convention in October 1971, saying that the NASA solution was 'simply not feasible, either technically or economically'. Browne estimated that the retrofit of older aircraft with noise-suppressing nacelles would cost the airline industry $1 billion.

Many outside of the Administration disagreed with Browne's assertion. At a press conference following the conference, Matthias Lukens, former deputy director of the Port of New York Authority, said 'There are government reports that do not support what Chairman Browne indicated'. Graham Hill, an executive of British Airports Authority 'strongly disagreed with Mr. Brown and backed the retro-fit proposal'. Representatives of three noisy airports – Los Angeles, New York and Boston – also held a news conference after Browne's talk and 'contended that tests on muffled nacelles had shown that significant noise reductions could be made in the Boeing 707 and DC-8 jets… (they agreed) that modifications might be costly but said the cost could be met by slightly higher fares and possibly some form of Government assistance such as tax incentives'.

The next day, Shaffer wrote a memo to DOT Secretary John Volpe, suggesting that proponents of the retrofit nacelles were 'absolute extremists'. He indicated that he continued to oppose retrofitting older plane engines, preferring instead to wait for the quieter, new engines.

Congress responded with the Noise Control Act of 1972, which strengthened the Aircraft Noise Abatement Act of 1968. This legislation imposed criminal and civil penalties for willful, knowing violations of federal noise standards. Enforcement was transferred to the EPA Administrator, and citizen lawsuits against the FAA Administrator for any failure to perform duties were permitted.

Breakthrough

Eventually, the 1968 Aircraft Noise Abatement Act and the 1972 Noise Control Act resulted in new Federal Air Regulations (FAR) and International Commercial Aircraft Organization (ICAO) agreements to reduce airport noise. The FAA's FAR Part 36 and ICAO Annex 16 specified three stages or 'chapters' that nations must comply with to achieve significant aircraft engine noise reductions.

In the US, Stage One aircraft were phased out from 1985. These were primarily the first-generation large jets, the DC-8 and B-707. Airlines had four choices:

- mothball the noisy old aircraft and replace them with quieter, newer aircraft, the most expensive option;
- re-engine the old aircraft if possible, at a cost of up to $6million plus 8,000 work hours;
- modify the engine and aerodynamics but suffer a performance penalty, shorter range and reduced payload;
- install a Hush Kit at a cost of up to $3million per plane, the least expensive solution.

Stage Two compliance began in 1985. The Airport Noise and Capacity Act of 1990 requires the phase-out of Stage Two aircraft by year 2000. These included all JT3D and some JT8D engines, such as those on the B-727.

Stage Three compliance began in 1994 when one-quarter of a carrier's fleet had to be noise compliant. Full fleet compliance was mandated by year 2000 in the US and April 2002 for Europe. Aircraft included in this phase are the older planes plus the DC-9, B-737 and the BAC One-Eleven with a Rolls-Royce Speg engine. (Wegg, 1999) Stage Three is about half as noisy as Stage Two.

Hush Kit companies have emerged to develop the technology pioneered by NASA to absorb the sound of noisy engines. Burbank Aeronautical Corp II (BACII) has built Hush Kits for more than 70 DC-8s, and QTV did Hush Kits for their own Fine Air fleet and then others. ABS Partnership (Airborne Express, Burbank Aero and Sanfran Corp.) had orders for more than 500 DC-9s, especially for Northwest Airlines. Quiet Skies Inc. developed Stage Three Hush Kits for modified Boeing 707's involving 15 major assemblies and 2,500 parts. FedEx ordered 1,300 Hush Kits for their 727s, and Nordam sells Hush Kits for the B737. Both QTV and European Aviation are working on Hush Kits for the BAC One-Eleven (Wegg, 1999).

However, Hush Kits are not as popular in Europe. 'Acoustic trades' are not permitted in Europe under ICAO rules. In the US, if two of three noise measurements are met, an acoustic trade-off can be made for the third. (The three noise measurements include takeoff, approach and sideline values.) Because of higher population densities in Europe than in America, more people are affected by airport noise. Older American aircraft such as the DC-9, B-727 and B-737 that have been retrofitted with Hush Kits are louder than the (European) Airbus A300 (Field, 1999).

In 1999, the European Union (EU) passed a law to ban all aircraft retrofitted with Hush Kits effective April 1, 2001. In February 1999, Neil Kinnock, the EU Transport Commissioner had declared that the proposed ban on Hush Kit fitted aircraft would not violate any EU/US trade agreement. US Trade Representative Charlene Barshefsky warned that such a ban would have 'profound impact' on EU/US relations, and David Aaron, Undersecretary of Commerce for International Trade said the directive 'certainly won't go unanswered', hinting at a lifting of the noise exemption for the noisy Concorde (*Financial Times*, 1999a).

EU officials backed off and postponed their Hush Kit ban one year from their original April 1, 2001 deadline (*Financial Times*, 1999b). Both sides agreed to work on Stage Four noise levels and aircraft modification. The law was repealed on March 26, 2002 after much negotiation.

In order to maintain US preeminence in aerospace research and technology, NASA's Aerospace Technology Enterprise has adopted a plan to reduce aircraft noise even more. 'NASA's strategies for noise reduction focus on quieter airframes and engines, and on improved procedures such as glide slopes and flight paths that reduce ground-level noise.' By 2005 NASA hopes to demonstrate technologies to reduce noise impact by 5dB, double that by 2011 and double that again by 2025 (National Aeronautics and Space Administration, 2000).

In addition, an AeroAcoustics Research Consortium has been formed at Glenn Research Center in Cleveland. NASA is partnering with GE Aircraft Engines, Honeywell Engines & Systems and Boeing to find codes for engine noise prediction. The consortium is administered by the Ohio Aerospace Institute, involving Ohio universities and the US Air Force Wright-Patterson Laboratory. Other aeroacoustics research is going on at Stanford, Florida State, Boston University, Arizona, Florida Atlantic, MIT, VPI and LSU.

Currently hush-kit technology is the most economical and reliable solution available for making older aircraft quieter. More than likely, aircraft engines of the future will incorporate the sound absorbing principles and materials of the Hush Kit in better designs and engineering. In addition, public pressure may very well build to a point where airports will be forced to allow none but the quietest aircraft to load and takeoff. In that case, Hush Kits may be developed for even later model plane engines.

Former NASA Administrator James Fletcher suggested many years ago that the new motto for the agency should be 'Fly Quiet'. Today, given the opposition to Hush Kits on the European front and fed-up Americans who live at the end of the runway, that motto might read: 'Fly Quiet...or Else'.

References

Boswinkle, R.W. (1968) 'Opening Remarks,' NASA Acoustically Treated Nacelle Program Conference, Langley Research Center, Hampton.

Environmental Protection Agency (1973) *Noise Source Abatement and Cost Analysis Including Retrofitting*, Aircraft/Airport Noise Study Report, NTID 73.5, EPA, Washington, DC.

Federal Aviation Administration (1972) *Aviation Noise Effects*, NTIS, ADA-154313, FAA, Washington, DC.

Federal Aviation Administration (1985) *Aviation Noise Effects*, US Dept. of Commerce: Washington, Washington, DC.

Field, D. (1999) 'Northwest Refurbishing 150 Older DC-9s,' *USA Today*, October 5, p. 5B.

Financial Times (1999a) 'EU Refuses to Delay Hushkitted Aircraft Restrictions Despite US Plea', February 11.

Financial Times (1999b) 'EU Postpones Implementation of Legislation Banning Hushkits, Giving US More Time to Re-Sell Their Hushkitted Aircraft,' April 29.

Gillen, D. (2003) 'The Economics of Noise' In D.A. Hensher and K.J. Button (eds) *Handbook of Transport and the Environment*, Pergamon, Oxford.

Hoban, F.T., Hoffman, E. and Lawbaugh, W. (1997) *Where Do You Go After You've Been to the Moon?* Kreiger Publishing, Melbourne.

Kent, R.J. (1980) *Safe, Separated, and Soaring: A History of Civil Aviation Policy 1962-1972*, US Department of Transportation, Washington DC.

NASA News Release (1971) 'NASA Chief Welcomes Jet Noise Challenge,' 71-222, November 3, p. 2.

National Aeronautics and Space Administration (2000) *Strategic Plan 2000*, NASA, Washington DC.

Schomer and Associates (2001) *Assessment of Noise Annoyance*, Schomer and Associates, Champaign.

Van den Bergh, J.C.J.M., Button, K.J., Nijkamp, P. and Pepping, G.C. (1997) *Meta-Analysis in Environmental Economics*, Kluwer, Dortdrecht.

Wegg, J. (1999)'The Rush to Hush,' *Airways Magazine*, September, p. 33.

Chapter 6

US Activities to Reduce Launch Costs

Harold Miller

Introduction

The key to space access is to reduce launch cost to orbit. This belief is shared by all involved with technology, science, commercial and military activities in space. This study reviews launch cost reduction efforts and proposals that have occurred over the last 40 years. In spite of almost continuous attempts by government and private industry to reduce launch costs by orders of magnitude, they remain in the range of $5000-$12,000/pound to orbit. The most optimistic programs hoped to reduce launch costs to as little as $100/pound to orbit. The efforts range from initiatives of small private companies to congressional attempts to mandate a ten-fold reduction in launch cost. The programs and proposals are loosely organized chronologically. This review focuses on the most important efforts to reduce launch costs; including Expendable Launch Vehicles (ELVs) Reusable Launch Vehicles (RLVs) (including single-stage-to-orbit – SSTO), and the Space Shuttle. It is not a comprehensive review of every proposal and program to receive serious consideration during the last 30 to 40 years.

In the 1960s the US National Aeronautics and Space Administration (NASA) proposed a Reusable Orbital Transport (ROT) that would have a launch cost of about $100 per pound to low earth orbit. The aerospace industry and the government have never come close to achieving this goal. Current launch cost is orders of magnitude higher than this estimate ($10,000 per pound to orbit) and has stubbornly remained at this level for many years. After almost 40 years, the cost of space launches should have been substantially reduced. Some cost reductions have been achieved in technology areas where performance increases and improvements in manufacturing have occurred, but access to space remains expensive.

During one five year period, between 1987 and 1992 there were 867 launch vehicles studies (NASA Special Studies Division, 1994). Science, technology, commercial and military interests place significant demands on space transportation systems. Survivability, endurance, and robustness drive the national security space systems. In the commercial area, cost and reliability are primary concerns. With the variety of legitimate demands placed on space launches, it is difficult to find one solution to the problem of reducing launch cost.

Launch operations represent a significant percentage of the total launch cost and involve significant safety concerns that resist cost reduction. Launching an airplane, a truck or even a person in the family car has reached a reliability that requires only routine consideration (except for weather). The space launch industry has not come close to achieving this level of reliability or performance.

There has been nothing really new in the development of launch vehicles in the past 30 years. A fundamental technical requirement of a launch vehicle is that it must accelerate from 0 mph on the pad to an orbital velocity of over 17,000 mph in a few hundred miles. The vehicle must carry all of its fuel unless it can gather oxygen at low altitude. The need to carry all of the fuel required for the complete mission is a major obstacle in launch vehicle development. Some of the proposed new technology approaches attempt to reduce the fuel mass fraction (fuel weight as a percentage of total vehicle weight) during the first stage. This has not yet been successful. The proposed designs rely on similar technology; large cost reductions are forecast but not achieved. Some of the non-technical costs of launch vehicles are functions of government regulations (e.g., OSHA and Environmental Protection Agency), high insurance rates, and unions. One report notes that 'These are often costs that other, but not all, countries do not have to bear. It would be unrealistic and unwise to depart from our current policies and expect to achieve significant cost savings' (Sawyer, 1992).

Achieving a significantly cheaper (order of magnitude) launch system is unlikely with current technology. Reductions in the fuel mass fraction will be small as long as current fuels are used, with their energy density. However, if the energy density of the launch fuel could be raised by a one order of magnitude, large weights could be put in low-earth orbit (LEO) with a reusable launch vehicle.

Another impediment is the lack of a national space policy regarding low-cost access to space. The military does not need low-cost vehicles to accomplish its missions, and the science community sees launch vehicle development as a threat to their own program funding.

Reusable Orbiter Transport

In a confidential 1964 proposal to Congress (now declassified), NASA proposed to build a Reusable Orbiter Transport (ROT) (Office of Technology Assessment, 1989) that would bring the cost to launch per pound to orbit to \$100 (Associate Administrator's Advanced Study Review, 1964). The 1964 paper estimates costs per launch ranging from \$35,000 per pound to launch a Vanguard (50 lbs payload) to a projected \$100 per pound for a post Saturn (1,000,000 lbs payload) and Reusable Orbital Transport (8,500 lbs payload) vehicle.

Big Dumb Boosters

In contrast to the ROT, which would reduce average launch costs by reusing the launch vehicle, the Big Dumb Boosters (BDB) concept proposes to achieve lower launch costs by reducing the complexity of launch vehicles in order to use simpler, lower-cost and widely available technologies. However, this may diminish vehicle performance and increase size.

In industry, mass production is often used to reduce production cost. Some believe that mass production techniques could bring down booster cost dramatically, especially if applied to the Big Dumb Booster. An Office of Technology Assessment (OTA) (1989) study makes an analogy between launch vehicles and trucks and high performance sports cars. Launch vehicle requirements necessitate a racing car performance.

The problem is that a truck will not get a payload to orbit. The fuel mass fraction of a truck is around 10%. If boosters could be built with mass fractions of 10% then the comparisons might be valid. But if performance requires that each individual booster be built like a Swiss watch, it is difficult to see how mass production will bring down the cost. The question is throw weight per booster and the amount of material available to deal with. The question becomes how much more of the payload weight can be used for structures to obtain increased reliability and reduced operating cost.

Space Shuttle

In 1972, President Nixon announced the launch of the Space Shuttle program, describing it as an 'entirely new type of space transportation system' that would revolutionize access into near space, by 'routinizing' it (see Figure 6.1). The Shuttle was intended to replace all other launch vehicles except the very smallest and the very largest. In his concurrent announcement about the Shuttle, James Fletcher, NASA Administrator, said that the shuttle is the 'only meaningful new manned space program that can be accomplished on a modest budget.'

NASA described the shuttle as an airplane-like vehicle that could by used over and over to take satellites to orbit and bring them back for repair and reuse. In a statement that later would require a major reversal in policy, NASA said that although the Shuttle could perform manned missions in orbit, its principal use would be to conduct missions with unmanned satellites more efficiently and economically than by using launch vehicles that can only be used once.

Preliminary cost estimates in a 1972 draft study titled *Space Shuttle Economics*, from the papers of George Low (NASA Deputy Administrator between 1969 and 1976), were very optimistic (Low, 1972). The total investment required to develop the Shuttle and procure flight hardware and permanent facilities was estimated at approximately $7 billion (1972 dollars). Peak annual funding of $1.3 billion was estimated not to require an increase in NASA yearly budget level, based on the assumption that the Shuttle would carry over 70% of the US mission model. NASA was not alone in making optimistic assumptions about projected mission models. Estimated cost per pound to orbit in 1972 was priced at $118, based on a flight schedule of 60 flights/year. It was estimated that each Shuttle flight would cost $8million and would carry 65,000 lbs. In comparison, the Titan launch cost was $24million, carrying 27,000 lbs for a per pound cost of $900. High inflation during the 1970s and early 1980s caused the cost to orbit to escalate rapidly.

However, some still used the $100 per pound estimate that was hoped for in the late 1960s and early 1970s without considering inflation.

(*Source*: NASA History Office)

Figure 6.1. Shuttle being Returned to the Kennedy Space Center

Optimistic assumptions about mission models have been repeated many times in proposals for developing new launch vehicles; they have not been unique to the Shuttle.

By the middle of 1973 it was becoming apparent that the cost of the shuttle could not meet the then current advertised cost of $160 per pound to orbit. The Comptroller General (1973) estimated a cost of $3500 per pound (1973 dollars). If inflated, at approximately 7%, over the last three decades this price has held up. The 1986 decision to eliminate commercial satellite launches from the Shuttle schedule increased the cost of operation by reducing the launch rate. Minimal staff levels at the Kennedy Space Center (KSC) have created conditions where any small problem may cause a Shuttle launch delay. Delays in Shuttle launches occur almost as frequently now as during the 1960s. In 1996, Gary Payton, while director of the X-33 launch vehicle project, noted that 'Limited money during development is cited as a cause of high launch cost. The money was not there to make it more reusable' (Borenstein, 1996). Inadequate funding during the development phase has also contributed to higher operating cost.

The cancellation of programs to develop a viable SSTO, such as the X-33 and National Aerospace Plane (NASP), and other replacements for the Shuttle, has led NASA to plan up grades to the Shuttle system and launch operations to reduce operating cost. In 1997 NASA predicted that liquid-fueled flyback boosters and other hardware improvements would cut the cost of launching a pound of payload to low Earth orbit to $2500 (*Aerospace Daily*, 1997). It hoped that additional upgrades to the Shuttle would lower the cost to about $500/pound by 2007 (Eisele, 1997a). NASA also expects that improved launch control systems at KSC will reduce launch cost by 50%. Other improvements, including non-toxic fuels, a liquid flyback booster, and improved guidance systems, will be implemented during a four-phase upgrade. It remains to be seen if these upgrades can in fact reduce Shuttle cost to the level desired.

It is difficult to compare Shuttle and ELV launch costs because the former can provide many more services than an ELV – e.g., the ability to return objects from orbit. It is difficult to calculate the value-added obtained by having a crew available.

Some missions do not require the services of the crew. For other missions, such as flights to service the Hubble Space Telescope or to assemble the space station, the crew is essential. The premium for flying crew into space should be the cost difference between the same payload flying on an unmanned vehicle vs. on the shuttle. Since the Shuttle and Titan IV carry about the same payload and have about the same cost does this mean that the crew rides and services are free? If so, then the Shuttle offers much more reliability than any other space vehicle. This is because of the value added by the crew's ability to assure that a satellite is functioning as it is released.

It seems misleading that crew costs are calculated the same way as a communications satellite. There is a case to be made that astronauts should be valued more highly than ordinary payloads in the cargo bay. There is a need to find a new way to compare launch cost that includes evaluation of efficiency and reliability.

Most Shuttle missions are dedicated to large science and the International Space Station. This policy closes off the possibility of the Shuttle being available to commercial ventures that would support manufacturing in space, such as the Leasecraft project. Even if a commercial product could be developed for production on the Space Station, there may not be adequate transportation to service it. Adopting a policy of co-manifesting on the Shuttle would support a higher load fraction and a higher launch rate that would lower the launch cost. The 1986 decision to stop using the Shuttle to launch commercial satellites into space had the effect of increasing the average launch operations cost by reducing the launch rate.

After 30 years of operation, and a planned future of at least 10 more, the Shuttle stands alone as a long-term investment in space technology. As Forrest McCartney, former Kennedy Space Center Director, points out 'It's the only way to get anything back [from space]. It is a truly unique, flexible vehicle' (Borenstein, 1996).

Advanced Launch System

During the 1980s, the Air Force became interested in developing a new launch vehicle capable of carrying large payloads at low cost. The Air Force needed a rocket that could be used to assure space access for the Strategic Defense Initiative (SDI). In 1988, the government announced the Advanced Launch System (ALS), a program to develop a new booster that would lower the cost to orbit from $3600 to $400/pound. Air Force Col. George Hess (*Washington Post*, 1988) said 'If things go as planned, the changes will shave 90% off the $2,000 to $4,000 it now costs to put a pound of spacecraft into orbit.' The ALS would be jointly developed by NASA and the DOD, and paid for by the DOD.

The ALS would be designed to carry more than 100,000 lbs to low earth orbit and was expected to be ready by 1998. In 1987, the estimated potential demand for the SDI program was 5 to 10 times the annual 400,000 lbs placed in space at that time.

In July 1987, the Air Force awarded seven one-year, $5 million 'concept definition' contracts to explore the technology requirements of the ALS as an overall system (Dornheim, 1987). Rocco Petrone, Rockwell Space Transportation Systems Division's president said 'What I see new in this approach is to take all our knowledge, which we've applied so well to get capability, and now give equal weight to driving cost down so that performance is only one factor, not the factor.' Cost reductions of an order of magnitude (10%) were projected.

In fact, none of the contractors' studies proposed a new breakthrough technology as a way to achieve the desired capability and cost, but included an array of options.

Rockwell proposed using recoverable components to reduce costs. Martin Marietta considered an aluminum-lithium alloy in the cryogenic tanks to save weight. The Boeing Company concept included a fly-back booster and required efficient, high-pressure engines that can be removed routinely from the vehicle for maintenance away from the launch flow (*Aviation Week and Space Technology*, 1987). Other proposals included expendable and partially reusable launch vehicles, upgraded materials, and the use of existing engines and solid rockets.

However, the Office of Technology Assessment (OTA) (*Aerospace Daily*, 1988) disputed the feasibility of the proposals:

> Barring deployment of Strategic Defense Initiative (SDI) weapons and sensors or a sharply increased rate of civilian payloads, it's not likely that US launch costs will be cut to one-tenth of existing rates as has been claimed by NASA and the Department of Defense.

An article in *Aerospace Daily* asserted that reductions could be made by making launch operations more efficient or turning NASA's launch oversight duties over to the private sector, but that drastic cuts could be achieved only under highly limited conditions. In fact, NASA did turn over much of the launch operations to a private contractor in the mid-1990s, resulting in some cost reductions.

According to John Wormington, 'With $300 a pound to orbit a "rate-dependent variable" ALS would have to launch at least 2.5 million pounds a year to be cost effective' (Morring, 1989). This works out to launching 100,000 lbs payload every two weeks. He continues 'We went back and challenged…the original assumptions in the space program, one of which is weight is bad … If you start out with the idea that what you're after is low cost and not low weight, it turns out you can in fact spend weight and save money.'

By 1989 DOD announcements about the ALS were more pessimistic. Thomas Moorman, Director of Space and SDI programs stated that cost reductions from the ALS would probably be a three-to-five times reduction from the price of $3,600 per pound for a Titan launch (*Defense Daily*, 1989). He also noted that achieving this would require a very large payload and at least 6 to 7 launches/year. Moorman said more work would be needed in advanced structures to address problems with development of the heavy-lift vehicle. Richard C. Henry (1989) noted that the cost is not likely to come down because they were designed for performance first, reliability second and cost effectiveness last.

In late 1989, Congress cut funding for the ALS program to $150 million, for propulsion development. The early stages of the SDI program could be implemented with existing Titan and Atlas rockets. The perceived need for the ALS capabilities did match the likely development costs, and the future of SDI itself was already in doubt.

New Launch System

In July 1991, the Bush administration announced (*New York Times*, 1991) a program to develop a new family of rockets that would be called the 'new launch system' (NLS). This was instead of building additional space shuttles. The announcement followed the April 16, 1991 directive of the National Space Council to NASA and DOD (*NASA Fact Sheet*, 1991) to jointly develop and jointly fund the NLS.

The NLS would support medium-to-heavy-lift launch requirements using modular components, including a 'core' launcher that would 'be stronger and heavier than absolutely necessary so that its maximum payload will not stress the craft to is design limit.' Its booster engines would be recovered and reused. Highly automated ground operations would reduce the crew needed to maintain current vehicles. The NLS would have an average cost of $500 for a pound to LEO and a failure rate of 1%. It would reduce launch cost, improve operations, and evolve into a system that is more responsive and resilient that system currently in use.

The estimated cost to develop the NLS was between $10 and $12 billion. Initial launch capability was expected in year 2000. Twenty years earlier, the Shuttle was developed with a 1972 dollar cost of over $7billion. It seemed unlikely that a comparable launcher could be developed at approximately one fifth the cost of the shuttle after allowing for inflation. In October, 1992, Congress voted against funding the NLS.

Spacelifter

In November, 1992, the National Space Council proposed that the Air Force assume sole responsibility for developing a new all-purpose national launch system called the Spacelifter (Sawyer, 1992b). Congress and the executive branch had been unable to agree on an approach for achieving a cheaper, safer, more reliable ride to orbit (Sawyer, 1992a). The Spacelifter proposal also called for phasing out the Shuttle at the earliest opportunity. The proposed 'new' launch system would be a clean break from the troubled old 'New Launch System (NLS)' program that the National Space Council (NSC) had supported starting in 1990 but which Congress refused to fund.

Nevertheless, the press reported that 'Several sources said the proposed system would not reduce launch cost enough to justify the estimated investment of from $5B to $7B and they noted that launches are based on 'wildly optimistic' estimates of launches per year.' A congressional aide, unnamed, said all the launch studies make 'naïve assumptions' and lack fundamental information.

Differing needs among the space community were apparent during the debate about the Spacelifter proposal. Edward Aldridge, chair of the Advisory Board, former secretary of the Air Force and president of The Aerospace Corporation of El Segundo, Calif. said that 'Without the introduction of a new launch vehicle that meets cost and performance goals, we can write off future US competitiveness in the area.' John Pike, director of space policy for the Federation of American Scientists, said the proposal,

> seems a lot like the NLS repackaged... With only a limited amount of money available, it might be better spent in upgrading existing rockets and developing a new crew capsule that can be used on a Titan Rocket to reduce the use of the space shuttle (Leary, 1992).

The views of Aldridge and Pike represent the differing priorities within the space community. Aldridge represented the Air Force view, while Pike's comment reflected concerns among the science community that funds for science application missions not be drained away for launch vehicle technology.

In the fall of 1993, Congress voted against funding the Spacelifter program. However, during the same session Congress allocated $10 million to the Advanced Research Projects Agency (ARPA) to investigate 'a new affordable, medium launch vehicle' in 1994 (Iannotta, 1993). Congress had been impressed with ARPA's quick development of the Pegasus rocket. However, the ARPA Launch Vehicle project was also not funded after the initial study.

Single Stage to Orbit

There have been several major programs attempting to replace the Space Shuttle. The combination of expendable rockets and a reusable orbiter, with a single-stage-to-orbit vehicle (SSTO) is not seen as necessarily the most efficient way to meet the US's future space requirements.

National Aerospace Plane/Reusable Aerospace Vehicle

The NASP was a joint DOD and NASA program, with Air Force and Navy participation, initiated in 1986. Over $ 2 billion was invested in the NASP program before it was cancelled in 1994. Although a flight demonstrator was never built, the program did develop and test numerous concepts and materials and NASA continues to maintain the NASP design concept. The NASP aerospaceplane concept included using an air-breathing engine (Figure 6.2).

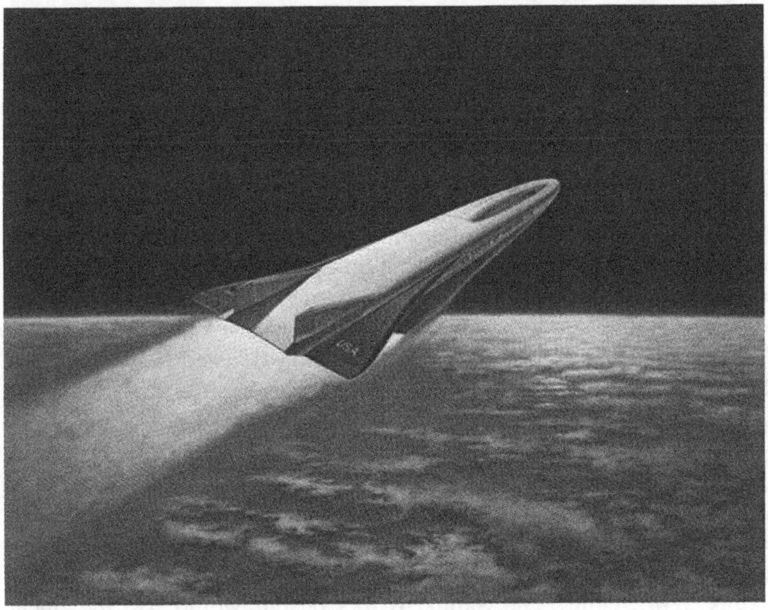

(*Source*: NASA History Office)

Figure 6.2. Image of the National Aerospace Plane

Boeing's Reusable Aerospace Vehicle (RASV) would have used technology based on the NASP (Sponable, 1990). The RASV would need to carry very dense and heavy liquid oxygen (Slush) for the rockets. The concept depends upon increased fuel/oxidizer energy density and also a sled to launch the first stage, which further reduces the fuel mass fraction.

The Air Force turned down a Boeing proposal to build an experimental half scale RASV prototype. NASA later bought into this idea with the current attempt to develop a SSTO. In addition, the vehicle structure depended upon very advanced materials such as titanium aluminide (Ti-Al) alloys, titanium aluminide metal matrix composites (TI-Al MMC), and/or carbon-carbon (C-C) refractory composites. The overriding idea in the Boeing proposal was to reduce the weight of the structure to make room for a payload or even fly.

Venture Star – X-33

In August 1994, President Clinton signed a space policy document requiring the Administrator of NASA to submit a plan for developing a reusable launch vehicle (RLV) by November 5, 1994 (*Space News*, 1994). He also called for the US Defense Department to improve the US fleet of expendable launch vehicles in the hope of making modest reductions in the cost of launching military satellites. NASA responded with a plan to develop a test vehicle called the X-33 that would be based on prior NASA technology research. The new vehicle, called the VentureStar – X-33, would replace the Shuttle, with first test flights planned before the year 2000.

In 1995, NASA launched the first phase of the three phase X-33 program. Phase II would then begin in 1996 with the selection of a single contractor. Phase III was scheduled to start in 1999 with full-scale production of the reusable launch vehicle (RLV) using private funding. The goal again was to cut launch cost dramatically, from $5,000-$12,000/pound to $1,000/pound to LEO. A half scale version of a RLV named the VentureStar (X-33) would be used to demonstrate advanced technologies that could dramatically increase reliability and lower the cost of putting payloads into space (see also Chapter 8 in this volume).

In 1996, NASA selected Lockheed Martin to build the VentureStar test vehicle. NASA would fund most of the program with an investment of $941 million; Lockheed Martin would invest $220 million. NASA would develop the technology for the half scale vehicle and then private industry would build the vehicle (Cast, and Amatore, 1996). The VentureStar was based on earlier work done by Lockheed's Skunk Works unit. The Skunk Works design would use the linear plugged nozzle engine that was designed in the 1960s for as an alternative to the bell nozzle engine (Fulghum, 1993).

Lockheed estimated that it could build an operational re-useable rocket for between $5 billion and $8 billion that would be able to fly 30 to 40 payloads a year. Lockheed Martin submitted a plan to NASA for a three-ship operational fleet that it felt it could operate profitably. The RLV would become operational around 2006 or 2007 – just in time to start phasing out the Space Shuttle.

Norman Augustine, CEO of Martin Marietta, was skeptical about the NASA plan, questioning whether the government would provide the level of funding needed to develop a 'breakthrough' technology (*Aerospace Daily*, 1994).

Some questioned the premise than an RLV can really be genuinely 'reusable'. Book (1999) notes that reusable systems are not fully reusable. Even with a high (0.99%) probability of recovery, he notes it could take 7 vehicles to maintain a fleet of 4 after 100 launches. Delta, the most reliable launch vehicle in history, has only achieved a success rate of 98% after 30 years. The success rate of more typical unmanned vehicles is 90-92%. Reusable vehicles must not only survive, but they must be in a condition to be refurbished which not always easy to achieve after a stressful launch.

However, Charles Conrad (1993), Vice President for New Business Space Systems at McDonnell Douglas Aerospace, and also a former Astronaut, in a 1993 response to a letter written by Alan Shepard, stated 'Studies show that no new technologies are involved (in SSTO development), nor are the engines required to run at as high a specific impulse or pressure as the shuttle engines.'

His view represented the general belief that the SSTO was a viable technology that could be achieved within a few years. Conrad's letter projected putting 20,000 pounds into orbit for $5 milion to $10 million ($500 per pound).

Gary Payton, NASA Director of the X-33 Program, believed that 'The program (SSTO) is to provide our nation with reliable, affordable means of access to space.' (Borenstein, 1996) However, by 1999 with mounting schedule delays, technical challenges and increasing budgetary problems, the X-33 seemed to be relegated to an advanced technology development with little prospect that industry would develop the X-33 into a commercial vehicle. By 2000, NASA had started a new program called the 'Space Launch Initiative;' the VentureStar program was cancelled in March, 2001.

Space Ship eXperimental – SSX / Delta Clipper – DCX

The SSX, proposed in 1989, was conceived as a SSTO rocket vehicle (without wings), that would have a vertical launch and landing, be fully reusable, and have a turnaround time of only two or three days between flights. The SSX concept was of great interest to the Strategic Defense Initiative Organization (SDIO).

> 'The SSX has a vast potential for both military and commercial uses in space' according to Daniel Graham. He believed that SSX technology would support the 'Brilliant Pebbles' plan to launch thousands of satellites as part of a space-based defense against nuclear missiles, and could 'in addition to strengthening Western security ... save the taxpayer half a trillion dollars over the next 10-15 years' (Mohr, 1989).

Reusability and short turnaround time would be the key to launch cost reduction.

Hoeser (1990), stated 'A key factor in any SSTO (Single Stage To Orbit) cost is the flight rate per year.' He based his economic projections on a nominal launch rate of once every two days as a norm for this system (SSX), using launch rates of up to 80 to compare cost with the shuttle launches of 10 per year.

The SSX program was funded, with implementation, to take place in three phases. In 1991, Phase II, McDonnell Douglas was awarded a contract to build the SSX test vehicle, which it renamed the DCX for Delta Clipper Experimental (Figure 6.3). The DCX was a one-third scale model of the proposed vehicle.

(*Source*: NASA History Office)

Figure 6.3. Image of the Delta Clipper Experimental Rocket (DC-X)

The first test flight in August 1993 the DCX reached an altitude of 46 meters. It completed two more flights before the SDIO program ended. However, additional funding was later obtained, and several more flights were made. Eventually the DCX was modified and tested as the DCX-A (advanced). It made four flights, eventually reaching an altitude of more than 3,000 meters, but after being seriously damaged in its final flight, the program was ended due to a lack of funding.

Shuttle-C

The Shuttle-C was proposed as an unmanned, heavy-lift launch vehicle that would be based on the Space Shuttle. Essentially the Shuttle orbiter would be replaced with an expendable cargo pod with two to three times the capacity of the existing Space Shuttle payload capacity. However the US Department of Defense was more interested in the NLS program. The Shuttle-C program was therefore not funded.

Assisted First Stage Launchers

Attempts to find ways to reduce launch costs have also involved efforts to develop ways to 'assist' first stage launchers, especially for smaller payloads. A number of options have been examined.

Rail Accelerator

As early as 1983, serious efforts to reduce fuel mass fraction were proposed by several governmental agencies. NASA's Lewis Research Center proposed an electromagnetic catapult called a rail accelerator (*Science Digest*, 1983). However, the proposal gave estimates of 20 to 30 years to build with a total development cost of $8B. In addition, the power required would be enormous – requiring the equivalent of the energy from all the US power plants for one second to launch a ton sized payload. Estimates of cost reduction are on the order of 5 to 50.

Electromagnetic Coil Gun

Sandia National Laboratory proposed an electromagnetic coil gun in 1991 (Rains, 1990). Miles Palmer, of Science Applications International Corp. concluded in a study for Sandia that 'The concept appears cost effective and if used to full capacity could reduce current launch cost 100 fold (initially of payloads of less than 100 kilo).' Launches could occur every 10 minutes. Cost per pound was cited as low as $20 to $50. The study concluded also that this was a very immature technology at this time. The cost would be $30M dollars to complete a demonstration project. High g-loads would preclude many payloads, including humans. Sandia Laboratories in New Mexico also predicted that a coil launcher that cost $1.2B could pay for itself in the first hour by sending an armada of satellites aloft. Using electromagnetic coil launchers the cost to put a pound into LEO would be about 1% of current rocket cost (*Washington Times*, 1991).

Light Gas Gun

In 1992, scientists at the Lawrence Livermore National Laboratories proposed a 'Light Gas Gun for Launch of Small Payloads' with an estimated launch cost of $600/lb (Henderson, 1990). The gun would be capable of 3,000 launches per year and would put about 22M lbs. into LEO during a 10-year period at a cost of $6 billion. It was hoped that a full-scale system could be in place by 1997. A principle use would be for the Brilliant Pebbles proposal for SDI.

MagLifter

The MagLifter is a form of electromagnetic propulsion proposed as a possible way to launch everything from 5,000 pound into orbit to SSTO type vehicles up to 200,000 pounds (David, 1995). This is another approach attempting to improve mass fraction to achieve a first stage boost to about 600 mph. Although Assisted

First Stage Launchers could eventually be cost effective, they have a high development cost and are restrictive in the payload that they can handle.

Other Launch Systems

Attempts to improve access to space by lowering launch cost have not been limited to NASA and Defense Department programs working with the major aerospace companies. Smaller companies in the aerospace industry have also looked for solutions to this problem, especially for smaller payloads.

Astroliner

In 1997, Kelly Space & Technology, Inc. of San Bernardino, California, proposed a reusable vehicle named the Astroliner (Cravotta, 1997). A Boeing 747 would tow the Astroliner like a glider to 40,000ft. where the Astroliner fires its rockets to ascend to a 75 mile altitude. The Astroliner's nose opens and it deploys a satellite, which is kicked into orbit. The piloted Astroliner would be able to land on any airfield. Horizontal payload integration would take only a few hours or at most a few days. This program is still in development.

Pathfinder

In 1997, Pioneer Rocketplane of Lakewood, Colo. announced plans to develop a piloted, air refueled space plane, the Pathfinder, capable of launching small- and medium-class payloads (Eisele, 1997). The Pathfinder would take off from a runway, firing its rocket after an air refueling at 30,000 feet. The rocket would propel the spaceplane to 70,000 feet where it would release its payload. Pioneer's president, Robert Zubrin said that 'We believe that Pathfinder can drop the cost of launching small and medium payloads to earth orbit by a factor of five compared to current cost with the same class booster; or by a factor of three compared to multiple payloads launched on the same vehicle.' The company believed an operational vehicle could be built for about \$100M. The approach uses a fully reusable first stage and an expendable second stage. Payload weights range from 990kg (polar) to 2475kgs 'due east.' This effort is no longer being pursued.

Star Booster

The Star Booster is a small, reusable rocket-powered booster designed to launch small satellites adapting existing technology by adding wings to a Zenit rocket. The first stage would be reusable after carrying a second and subsequent stages to mach 5.5 where they are launched. The Star Booster would return to earth under turbofan power like an airliner. The Zenit will be refurbished, refueled and launched many times. Cost per pound should be below the current SSTO goal of \$1000 a pound. The Air Force awarded a phase 1 design contract to Starcraft Boosters, Inc., headed by former astronaut Buzz Aldrin, in June, 2002. The company believes that it can preserve more than 70% of the dry vehicle weight. This poses a refurbishment problem and it makes the probability of a quick turn around more difficult.

Lockheed Martin Launch Vehicle (LMLV)

The Lockheed Martin Launch Vehicle (LMLV) system, which is actually a family of launch vehicles, is designed to provide a highly reliable, affordable launch and maintain an aggressive development schedule (McCurry and Carter, 1997). Three configurations are available to meet a wide range of payloads. Maximum use of exiting stage components, subsystems, and facilities is attempted to minimize development cost schedule, and programmatic risk.

Conclusions

During the last 30 to 40 years, there have been hundreds of conferences, papers, reviews and advisory groups studying a wide variety of approaches to reducing launch costs. Many attempts to develop some of the most promising concepts have failed. Although small advances have been made in specific areas of technology, there has not been a breakthrough in any aspect of launch costs.

Factors influencing launch costs are not only technical, but include: vehicle size, payload size, government regulations, including safety and environmental, insurance costs, labor costs, including union requirements, reliability and failure rates, reusability, fuel variables, materials used and many others.

It is difficult even to find agreement on defining launch costs, which reflects the lack of clear definitions of what goes into 'launch costs.' For example, the following definition of 'payload' is from the Commercial Space Act of 1998, '…the term 'payload' means anything that a person undertakes to transport to, from, or within outer space, or in sub-orbital trajectory, by means of a space transportation vehicle, but does not include the space transportation vehicle itself except for its components which are specifically designed or adapted for that payload.'

This definition can raise many questions (Office of the Associate Administrator for Commercial Space Transportation, 2003). For example, is the shroud part of the payload? During manned space flight does the orbiter constitute a payload? Its weight is a quarter of a million pounds. The payload in the cargo bay is only about 40,000 lbs unless going to higher orbits. In transportation systems with a mature technology, operations development cost and reliability are well understood.

But the problem would seem to extend well beyond this. The new world of public-private cooperation is a complex one. The traditions of NASA and many other government agencies were more akin to a planning bureaucracy than an undertaking that provides for customer demands. In the past, it was heavily funded by the government with its objectives were political rather than commercial with an element of pure scientific research included. While some countries manage to offer low cost space services (Russia can launch people at about $60 million a time and has experienced less accidents than the US), NASA because of its ingrained philosophies has been less able to adapt.

References

Aerospace Daily (1988) 'OTA Report Counters Claims of Ten-Fold Cut in Launch Costs,' September 15.

Aerospace Daily (1994) 'Augustine Skeptical about New Reusable Launcher Profitability,' November 3, p.180.

Aerospace Daily (1997) 'NASA Wants to Cut Shuttle Payload Launch Cost to $2500 a Pound,' April 7, p.38.

Associate Administrator's Advanced Study Review (1964) *Launch Vehicles*, NASA declassified document, June 26, NASA, Washington.

Aviation Week and Space Technology (1987) 'Boeing ALS Design includes Turbine, Rocket Powerplants,' September, p.34.

Book, S.A. (1999) 'Inventory Requirements for Reusable Launch Vehicles,' *Space Technology & Applications International Forum*, January 31-February, 4.

Borenstein, S. (1996) 'Costly History Dogs Shuttle on Anniversary,' *Orlando Sentinel*, July 12, p.A1.

Cast, J. and Amatore, D. (1996) 'Lockheed Martin Selected to Build X-33,' *NASA News Release*, July 2.

Comptroller General (1973) *Analysis of Cost Estimates For The Space Shuttle and Two Alternate Programs*, Report to Congress, Washington DC.

Conrad, C. (1993) 'Spaceflight: Yield to Superior Technology,' *Washington Post*, July 19, p.A14.

Cravotta, D. (1997) 'Airplane-Towed Space Vehicle to Launch Satellites,' *Final Frontier*, July/August, p.12.

David, L. (1995) 'NASA Studies Use of Magnetic Launch Pad,' *Space News*, September 19-25, p.6.

Defense Daily (1989) 'ALS Could Cut Lunch costs 3-5 Times,' May 3, p.188.

Dornheim, M.A. (1987) 'Air Force Awards Contracts for Phase 1 ALS Studies,' *Aviation Week and Space Technology*, July 20, p.25.

Eisele, A. (1997a) 'NASA Plans Four-Phased Shuttle System Upgrade,' *Space News*, April 7-13, p.11.

Eisele, A. (1997b) 'Pioneer Touts Space Plane a Way to Cut Costs,' *Space News*, February: 10-16, p.8.

Fulghum, D.A. (1993) 'Skunk Works Design May Cut Launch Cost,' *Aviation Week and Space Technology*, August 16, p.76.

Henderson, B.W. (1990) 'Livemore Proposes Light Gas Gun for Launch of Small Payloads,' *Aviation Week and Space Technology*, July 23, p.78/79.

Henry, R.C. (1989) 'Launches Into Low-Earth Orbit Should be Economical, Routine,' *Aviation Week and Space Technology*, November 27, p.93.

Hoeser, S.J. (1990) 'The Cost Impacts of True Spaceships,' *The Journal of Practical Applications in Space*, **1**(4), 1-38.

Iannotta, B. (1993) 'US Lawmakers Nix Spacelifter in Favor of ARPA Launch Vehicle,' *Space News*, November 15-18, p.3.

Leary, W.E. (1992) 'A New Breed of Rocket is Urged to reduce Space Program's Cost,' *Washington Post*, November 20, p.A18.

Low, G.M. (1972) *Space Shuttle Economics Simplified - DRAFT*, 'From the George M. Low papers,' NASA Washington DC.

McCurry, J.B. and Carter, W.P. (1997) *The Lockheed Martin Launch Vehicle (LMLV) Solution for Small Payloads*, IEEE Paper 0-7803-3741, July, p.343.

Mohr, H. (1989) 'Into the SSX Blue Yonder,' *Washington Times*, May 17, p. F4.

Morring, F. (1989) 'Advanced launch System Program Goal: Cut Payload Costs,' *Defense News*, January 9, pp.9-109.

NASA Fact Sheet (1991) 'New Launch System,' George C Marshall Space Flight Center, August 29.

NASA Special Studies Division (1994) *Access to Space: Issues As Reflected in Recent Unclassified US Studies*, NASA Special Studies Division, Washington DC.

New York Times (1991) 'US Plans to build no new Shuttles,' July 25, p.A18.

Nixon, R. (1972) 'Announcement on the Space Shuttle,' NASA History Office, Washington DC.

Office of Technology Assessment (1989) *Big Dumb Boosters: A Low Cost Space Transportation Option?* Office of Technology Assessment, Washington DC.

Office of the Associate Administrator for Commercial Space Transportation (2002) 2002 *Reusable Launch Vehicle Programs and Concepts*, Department of Transportation, Washington DC.

Rains, L. (1990) 'Funds Sought for Coil –Gun Device,' *Space News*, January 15-21, p.1/37.

Sawyer, K. (1992a) 'Shuttle Replacement Proposal Appears Stuck on Launch Pad,' *Washington Post*, November 21, p.A2.

Sawyer, K. (1992b) 'Space Panel to Call for Replacement of Shuttle,' *Washington Post*, November 19, p.A10.

Science Digest (1983) 'Riding a Rail Into Orbit,' December.

Space News (1994) 'Beware the Triple Threat,' September 19-25, p.14.

Sponable, J.M. (1990) 'Reliable Low Cost Space Transportation Impossible or Intolerable?' *The Journal of Practical Applications in Space*, **1**(4), 39-47.

Strobel, W. (1987) 'Airman Expects Routine Space Trips,' *Washington Times*, September 10, p.A11.

Washington Post (1988) 'Reagan Clears Space Booster For the 1990,' *Washington Post*, January 14.

Washington Times (1991), 'Coil Launcher May be Space Age's Future,' March 15, p. B7.

Chapter 7

Spaceports

John T. Sheahan and Francis T. Hoban

Introduction

A spaceport is a facility from which launch vehicles are sent into space, transporting the people and equipment needed to conduct commercial, scientific, and technological endeavors in orbit and beyond (Figure 7.1). Early 21st century spaceports are Earth-based facilities that launch space vehicles and their payloads into space and, if necessary, land them on return. They are similar to the modern airport in design and operation. Since airports generate impressive revenue and taxes and stimulate local economic development, it is easy to see how decision-makers might view spaceports as potentially attractive investments.

The commercial space business does have a significant economic impact today. The Federal Aviation Administration (FAA) (1971) reports that 'US economic activity linked to the commercial space industry in 1999 totaled over $61.3 billion. Commercial space transportation was directly and indirectly responsible for $16.4 billion in employee earnings in the United States. Over 497,000 people were employed in the United States as a direct or indirect result of commercial space transportation and related industries.' It seems clear that the commercial development of space transportation will also have a role in future economic growth (Federal Aviation Administration, 2001).

These data do not tell the whole story. Space flight remains a high-risk, high-cost endeavor. The world's air transportation system is mature, routine, and safe. Enormous infrastructure investments have been made over decades to bring the system to this point. A contemporary airport can process about 750 flights a day, moving thousands of passengers and crews, and tons of cargo. Airlines and airport operators profit from this high demand. In contrast, the demand for commercial access to space is low, and the high cost of processing and launching payloads will continue to suppress demand for the foreseeable future. Additionally, a number of technical, safety, and economic obstacles in the commercial space industry must be overcome before spaceports can begin to achieve financial success. Currently, investment decisions in spaceports may be prompted by national, regional, or state pride, or by the desire to maintain technical competence but they will need subsidies until the commercial space industry develops sufficiently.

Today, almost two dozen active spaceports are capable of launching payloads into orbit; two of them have launched people. Of the four US civil spaceports licensed by the FAA, three have launched commercial payloads: Spaceport Florida, California Spaceport, and the Kodiak Launch Complex in Alaska. The fourth spaceport is the Virginia Space Flight Center in Wallops Island, Virginia. Further, the FAA Office of Commercial Space Development estimates that in the next decade, an annual average of 46 commercial payloads will be flown on 32 launches worldwide (Federal Aviation Administration and Commercial Space Transportation Advisory Committee, 2002). This is not enough to profitably support current US civil spaceports, let alone new ones.

(*Source*: NASA History Office)

Figure 7.1. View of John F. Kennedy Space Center from the Shuttle Atlantis

The objective in this chapter is to look at the nature of spaceports and to consider the ways in which they may develop in various circumstances. The role of public policy, at least into the immediate future would seem central given the economic characteristics of spaceports. But ultimately they will require commercial support which itself will be, at least partly, determined by official attitudes. The aim is to look into the next twenty years or so to see the challenges confronting spaceport policy making. Longer assessments may be needed, and may be important, but they require more in depth analysis spaceports (George Mason University NASA Continuing Career Program, 2002).

Methods of Ownership and Operation

A spaceport is a special case in the family of port infrastructure facilities. Much of its traffic is unidirectional; it supports a specialized business niche; its payloads, launch vehicles, and equipment are very costly; and environmental and safety considerations are paramount.

A spaceport includes the facilities, equipment, personnel, and vicinity required to prepare a spacecraft for flight, initiate and manage the flight, and perhaps receive the craft at the end of the flight. For Earth-based spaceports, vicinity refers to the land or sea occupied by the facilities and equipment. For space-based spaceports, vicinity refers to the orbit and operations envelope around the spaceport. Unlike an airport, a spaceport can be dispersed over several locations, including down-range instrumentation facilities and space-based communications equipment (Vision Spaceport Partnership, 2000). For example, US abort landing sites are unique to the Kennedy Space Center and are not typical of spaceports.

A spaceport shares some basic characteristics of airports and seaports. But is has the exception that its activities affects a wider range of air and land corridors during launch and landing. Also, spaceports are more complex than airports because of the high-energy propellants, narrower safety margins, and extremely complicated and redundant systems they require.

Spaceports accept payloads from a customer, integrate payloads into a launch vehicle, prepare vehicles for flight, and launch vehicles and their payloads. Once in orbit, the payload either permanently deploys into space, as is the case for communications satellites, or it performs experiments and sends data back to Earth. A payload designed to be returned to the customer after flight is flown on a reusable launch vehicle (RLVs).

The US Department of Transportation (DOT) licenses commercial spaceports. To date, the body of specific laws, regulations, and procedures applicable to spaceports is less comprehensive than for airports and seaports. As a result, spaceport operators have more autonomy in formulating unique procedures for internal processing and daily operations. The enabling authority for funding and operating commercial launch activities derives from legislation enacted by the states in which the spaceports are located. Some daily operations are governed by federal laws and regulations generic to all industrial functions, such as the Occupational Safety and Health Act.

As commercial demand for launch services grows, and consequently the number of launch sites proliferate, spaceports could be based in regions rather than in individual states. Adjacent states would have a vital interest in the authority and operation of a regional spaceport that would serve the transportation needs of its citizens and businesses. However, officials thus far have been unwilling to commit their state's resources to a spaceport located in another state. Consequently, most US civil spaceports are likely to continue to be owned and operated by authorities, commissions, or boards established at the state or regional level. This ownership concept has worked well for large, city-specific or regional airport complexes and for the three US spaceports with orbital capabilities. But there are a number of other models.

Private Ownership/Fixed Launch Site

A developer of a single-class launch vehicle might choose to develop and operate a spaceport. This option has distinct advantages, including the ability to select an optimal location to enhance launch system performance and reduce costs. The owner/operator also exercises complete control of scheduling and is relatively free of outside influences. The ability to operate under fewer US Government rules and regulations may make this an attractive alternative despite the potentially high development costs.

Private Ownership/Mobile Launch Site

Pegasus and Sea Launch are the existing mobile launch systems. Pegasus is an air-launched vehicle (taken aloft and released) and can carry a payload of up to 1,000 lbs to low-Earth orbit. Sea Launch is a private mobile spaceport that uses a converted oil-rig platform at an equatorial launch site. Its ocean-based launch services provide commercial satellite customers with a direct route to geosynchronous transfer orbit. It can lift a spacecraft of more than 12,000 lbs or place a payload into a higher perigee, helping satellite operators attain a longer satellite service capability.

Sea Launch's use of marine operations reduces launch infrastructure and minimizes operational costs. Another potential mobile launch system is the Boeing Air Launch System, which would use a modified 747 aircraft to carry a launch vehicle to high altitude before release.

Public Ownership/Shared Facilities

The Florida Space Authority, Virginia Commercial Space Flight Authority, and California Spaceport use the public ownership/shared facilities model. The federal government owns many of the launch preparation and operations facilities but shares them with the civil operator, who may have to provide a launch stand and associated equipment. This is advantageous in that it provides expensive and often unique facilities and launch personnel at an incremental cost. However, many customers perceive the disadvantages of a shared facility to be uncertain scheduling and mandatory compliance with a government bureaucracy.

State Ownership/Federal Government Financing

The Kodiak Launch Complex is owned and operated by an authority established by the state of Alaska and funded by the federal government. The complex offers good port facilities, a regional airport, and an adequate surface transportation systems. The island also has air and sea shipment transfer capability available at Anchorage, 250 miles to the north. This model is difficult to replicate because of the investment needed to design and construct the facility – the federal government provided the necessary $26 million investment.

Federal Government Ownership/Contractor Operations

The federal government could fund the development of a spaceport and lease the operations to a launch vehicle developer, private corporation, or state or regional authority. This is similar to the expansion of airport capabilities, especially in the construction of runways and taxiways. An airport trust fund largely supports such expansion, and is supplemented by the passenger facility charges paid by airport users.

Independent Public Authority

Taxpayers often view an independent public authority favorably. The Port Authority of New York/New Jersey is an example of a financially independent, self-supporting public agency that receives no tax revenues from any state or local jurisdiction and has no power to tax. The Port Authority relies almost entirely on revenues generated by users through tolls, fees and rents.

Foreign Trade Zone

Major hub airports and seaports such as Orlando International Airport and Port Canaveral enjoy a commercial advantage known as a Foreign Trade Zone (FTZ).

Congress created the FTZ authority to permit foreign and domestic merchandise to be admitted into the US without formal customs entry or payment of duties. Advantages of locating in a FTZ include deferred warehousing, distribution costs, and duties; reduced assembling, processing, manipulating or manufacturing costs, and duties; and elimination of costs and duties in the event of export or damage. British, French, Italian, German, and Japanese aerospace companies now send communications satellites and components to Port Canaveral for processing and launch.

Spaceport Design Considerations

Safety is always the most important factor in spaceport design, as increasingly are environmental factors. But a spaceport will only succeed by meeting its customers' needs for payload processing and launch services in a safe, reliable, cost-competitive, and user-friendly manner. A number of factors dictate spaceport size and design costs (George Mason University NASA Continuing Career Program, 2002).

Location

A spaceport requires a clear down-range area where spent launch stages and/or aborted launch vehicles can return to Earth without injuring people or property. Range limitations are usually expressed in terms of launch azimuth – the range safety limits or the ground tracks of a launch vehicle that do not traverse populous land masses – and distance down range. In most cases, a payload customer will require access to the payload in orbit through the use of available ground facilities. Azimuth constraints at the spaceport will determine whether the payload can be delivered to an orbital inclination with a ground track that reaches a latitude acceptable to the customer.

A spaceport must be located to maximize efficiency in delivering a payload mass into orbit. The orbit inclination of most US spacecraft is about 30 degrees, which is where the orbital plane intersects the equatorial plane at 30 degrees. This is why the current predominant launch site is Cape Canaveral at 28.5 degrees north latitude. A due east launch adds the Earth's surface rotational speed to the horizontal velocity acquired by that launch vehicle.

Changing the launch azimuth of the launch vehicles can result in higher orbital inclinations, but with a commensurate reduction in the velocity increment from the Earth's rotation and a related reduction in payload capacity. Achieving an orbital inclination lower than the latitude of the launch site requires orbital plane change maneuvers and additional performance from the launch vehicle.

Because demand is high for the delivery of payloads to geostationary or geosynchronous orbits in the equatorial plane, a launch capability at or near the Earth's equator is favorable. A site at that latitude takes full advantage of the Earth's surface rotational velocity at that inclination. Indeed, the three major orbit inclinations have different economic potentials.

- Geosynchronous orbits can be at any Earth inclination and are usually at altitudes where the orbit period directly relates to the period of the Earth's rotation in which they may have a repeating ground track. Communication with geosynchronous payloads requires steerable tracking antennas at the ground stations.
- Geostationary orbits are particular types of geosynchronous orbits in which the orbit inclination coincides with the Earth's equatorial plane and the orbit period coincides with the Earth's rotation rate. The ground track is a fixed point on the Earth's surface at the equator. Communication with geostationary payloads requires only fixed-orientation antennas at ground stations.
- Polar orbits at any altitude are in a plane perpendicular to the Earth's equator and pass over the North Pole and South Pole. A polar-orbiting satellite can scan the entire surface of the Earth in one 24-hour period. Satellites in low-altitude polar orbits can obtain enhanced resolution of Earth observations for resource and weather information.

Market Decisions

Spaceport developers must decide what part of the commercial space industry to target in their strategic goals – commercial space market segments have different operational requirements. Specific operations will in turn determine the type, size, and possibly the location of the spaceport.

A spaceport launching crews and passengers into orbit requires a more complex infrastructure than a spaceport specializing in robotic launches. This means consideration of:

- Launch vehicle sizes and types (expendable or reusable);
- Payload size, weight, orbit, power, and environmental resources;
- Payload operations and/or deployment on-orbit;
- Crew and passenger versus robotic missions;
- Vertical and/or horizontal launch and landing.

Spaceport developers intending to offer multiple services will have to evaluate parallel site development or, more likely, evaluate a site's development over several years. It is not sufficient simply to identify a parcel of available real estate as a future spaceport and expect financiers, entrepreneurs, and prospective customers to commit their resources. The decision to develop a site must be market driven; otherwise, the spaceport risks becoming a burden on taxpayers or other investors.

Public Interest

A spaceport authority is responsible for minimizing potentially adverse effects on the physical environment surrounding the spaceport and for protecting people in the local and regional area from potential launch malfunctions and other risks. Spaceport developers should plan and implement these actions in consultation with appropriate officials of local, regional, state, and Federal Governments; the interested public; and other stakeholders (George Mason University NASA Continuing Career Program, 2002).

Spaceport Authority

A spaceport authority has ultimate control over all spaceport operations, similar to the control of airport operations by airport authorities. It resolves any conflicts of use and determines priorities for launch schedules, range use, and other facilities uses. It also provides the land and the infrastructure necessary to launch payloads with the associated launch systems. Security, safety, fire, general utilities, water, electricity, and propellants and gas handling are some of the basic services provided, along with separate generic launch pads and runways, if necessary. Spacelines may have the option, with suitable planning, to provide vehicle-specific launch sites or pads on the spaceport facility.

Spacelines

Spacelines is a term used in the same context as airlines. Spacelines one day will be the interface between spaceport customers and spaceport operations. Spaceport developers need to understand these roles, as well as the differences among various spacelines and how they might use the spaceport. A spaceline will be a client of the spaceport and provide services to its own customers. In 2001, launch vehicle manufacturers served as spaceline companies. Future spacelines may purchase launch vehicles directly from manufacturers and offer space launch services to payload customers using spaceport facilities. Payload preparation and integration will continue to be performed by payload and launch vehicle personnel, who are the most knowledgeable of specific launch vehicles and systems.

Supplying Spaceport Services

The federal government has operated US launch ranges for over 50 years. In the early years, military and civil government agencies conducted most of the launches. In recent years commercial operations have increased from the four commercial facilities. The first commercial launch was the lunar Prospector mission flown from Florida Space Authority's launch complex in January 1998. The current and potential future facilities differ considerably in their features and some of these are outlined below (see also, Federal Aviation Administration and Commercial Space Transportation Advisory Committee, 2001).

Existing US Facilities

Cape Canaveral Spaceport is a partnership among the US Air Force's 45th Space Wing at Cape Canaveral Air Force Station, Kennedy Space Center, and the Florida Space Authority. It is the world's busiest and most competitive launch site and can support a variety of launch trajectories and handle the smallest suborbital experiment payloads up to the largest commercial satellites, and defense and space exploration missions. Six active pads are used to launch Delta, Atlas, Titan, and Athena launch vehicles. Kennedy Space Center has two pads for the Space Shuttle.

In addition, the Spaceport offers payload processing facilities, security services, and access to the world's only quadramodal transportation system: space, rail, water, and road transit. Port Canaveral maintains a Foreign Trade Zone.

Cape Canaveral Air Force Station is adjacent to Kennedy Space Center and the 45th Space Wing, US Air Force, is responsible for the daily operations. The Eastern Test Range extends 10,000 miles into the Atlantic Ocean. In addition, the 45th Space Wing oversees protection of the environment and has responsibilities for the preservation and protection of historical artifacts. The 45th Wing commander or his deputy serves as on-scene commander and provides disaster response forces for all space launches and all hazardous materials operations.

Kennedy Space Center is the NASA center of excellence for launch and payload processing systems. It handles the checkout, launch, and landing of the Space Shuttle and its payloads. In addition to supporting crewed missions, it also coordinates all launches of NASA payloads and expendable launch vehicles (ELVs), whether those launches take place at Cape Canaveral Spaceport or elsewhere. In cooperation with the State of Florida, it is developing a space commerce park dedicated to space research. Construction on the Space Experiment Research and Processing Laboratory, dedicated to life science payloads and research for the International Space Station, began in 2001.

Florida Space Authority is responsible for statewide space-related economic and academic development, including regulatory and operational support to the space transportation industry. The authority operates like an airport authority, providing infrastructure, access, and operational support for expendable reusable and suborbital launch vehicle programs. It currently manages two operational pads and a state-of-the-art Space Operations Control Center at Cape Canaveral Spaceport. The authority provides financing, advocacy, technical support, business incentives, and facility/infrastructure development for space-related projects.

Vandenberg Air Force Base is on the central coast of California and is the US's premier polar launch site with five launch complexes, and one commercial spaceport. It is the only military installation in the US that launches robotic government and commercial satellites into polar orbit. The base serves as the headquarters for the 14th Air Force, which is responsible for all US Air Force space programs and on-orbit space operations worldwide, and is home to the 30th Space Wing, which manages and supports spacelift operations and flight tests of the US's intercontinental ballistic missile force. It operates the western range network, a geographic region consisting of instrumentation sites along the California coast for use by the US Government and commercial firms operating from Vandenberg.

California Spaceport, located near Lompoc, California, is a commercial launch services company operated and managed by Spaceport Systems International, LP, a limited partnership between ITT Federal Services Corporation and California Commercial Spaceport, Inc. It was the first commercial launch site to be licensed by the DoT Office of Commercial Space Transportation, in 1996. Spaceport Systems International has signed a 25-year lease with Vandenberg Air Force Base to use a commercial launch site. In 2000, California Spaceport conducted its first two orbital launches on a Minotaur launch vehicle.

Wallops Flight Facility NASA has operated a sounding rocket range at Wallops Island, Virginia, since 1945. NASA conducted its first orbital launch from a balloon used to study atmospheric density at the facility. Since that time, several orbital and suborbital missions have been conducted at Wallops. It has five launch areas, including one for heavy-lift suborbital rockets and one for classified payloads, two blockhouses, and preparation facilities are operational. Wallops Flight Facility has conducted more than 1,500 launches since 1945 and its personnel and mobile range equipment participate in about 20 to 35 suborbital launches per year from Virginia, Alaska, and other sites such as those in Australia, Norway, and Sweden.

Virginia Space Flight Center is a partnership with the Virginia Commercial Space Flight Authority, NASA, the Virginia Economic Development Partnership, Old Dominion University, Virginia's Center for Innovative Technology, and DynSpace, LLC. It develops and operates a multi-user spaceport at the Wallops Field facility that provides low-cost, safe, reliable, schedule-friendly space access to commercial, government, and academic users. Its capabilities include suborbital and orbital missions with payloads of up to 8,500 lbs. Virginia Space Flight Center leases a variety of facilities from NASA including a launch range with fixed and mobile tracking, telemetry, and command destruct capability, a modern range control center, and versatile payload processing and integration facilities.

Kodiak Launch Complex is the first US launch site not collocated with federally owned real estate. It is on Kodiak Island, Alaska and facilities include a launch pad and service structure, an integration and processing facility, a spacecraft assembly transfer facility, a payload processing facility, and launch control center. It was built by the state of Alaska to serve commercial enterprises launching telecommunication, remote sensing, space science, and DoD satellites.

To develop and promote the space industry in Alaska, the state established the Alaska Aerospace Development Corporation. This is a public corporation that operates administratively within Alaska's Department of Community and Economic Development and is affiliated with the University of Alaska. It is part of an integrated land, sea and air transportation system. The site is a FTZ and one attractive feature for potential developers is the absence of local competition for launch range services.

Edwards Air Force Base in Mojave California is NASA's Dryden Flight Research Center and the Air Force Flight Test Center. Edwards is a backup site for Space Shuttle landings. It has no vertical launch facilities but it has been used as a site for reusable launch vehicle demonstration programs.

White Sands Missile Range is located near Las Cruces, New Mexico and includes the NASA White Sands Flight Test Center. It is operated by the US Army and is used mainly for launching sounding rockets. The site also supports Ballistic Missile Defense Organization flight-testing and is used as a test center for rocket engines and experimental spacecraft. Facilities at White Sands include seven engine test stands and precision cleaning facilities. White Sands is also a Space Shuttle's alternate landing site.

Sea Launch is the world's only ocean-based launch services company. It provides commercial satellite customers with a direct and cost-effective route to geosynchronous transfer orbit. From its equatorial launch site, the Sea Launch Zenit-3SL rocket can lift a heavier spacecraft mass or place a payload into a higher perigee, helping satellite operators attain extended satellite service. Boeing, a 40% partner of Sea Launch, manufactures the payload fairing, is responsible for analytical and physical spacecraft integration, and manages overall mission operations, including the home port in Long Beach, California.

Proposed US Spaceports

In addition to its existing launch capacity, the US at both the federal and state level, has been active in considering the development of new capacity. These are various stages of planning, and are seen as serving a wide range of potential markets.

National Coalition of Spaceport States was established in February 2001 by 14 states (Alabama, Alaska, California, Florida, Montana, Nevada, New Mexico,

Oklahoma, South Dakota, Texas, Utah, Virginia, Washington, and Wisconsin) with the purpose of uniting behind a mission of regaining America's predominance in commercial space development. It hopes to influence commercial space policy, with an emphasis on spaceports and has goals that include the definition, development, and operation of spaceports in the US with an emphasis on promoting the development of new spaceports in Coalition states.

Alabama Spaceport. The Alabama Commission on Aerospace has proposed Baldwin County, across the bay from Mobile, as a site for a spaceport. Its goal is be operational in 10 years to service third generation RLVs.

Mojave Civilian Test Flight Center. The Mojave Airport was established in 1935 as a county facility with taxiways and basic support facilities suitable for general aviation. Later, the it was taken over by the federal government and turned into a Marine Corps auxiliary air station. In 1961 it was reacquired by the county government and turned into a civilian flight test center. It now has four rocket test sites, runways, test stands, and engineering buildings.

Montana Spaceport. The state of Montana proposed to fly RLVs from a site near Great Falls and Malmstrom Air Force Base and is in the process of obtaining a commercial spaceport license from the FAA. However, since the cancellation of the X-33 program, the state has suspended development plans.

Nevada Test Site. Kistler Aerospace selected the Nevada Test Site northwest of Las Vegas, as a spaceport for the K-1. Although it does not have any launch infrastructure, the Nevada Test Site has existing basic infrastructure that can be used for support facilities such as a paved runway, water, roads, and power.

Oklahoma Spaceport. The state of Oklahoma designated the Clinton-Sherman Air Force Base at Burns Flat as a future spaceport. In addition to the existing infrastructure, there are plans to build a rocket engine manufacturing facility. No money has been allocated for development, but the designation allows continued spaceport planning and development when funding sources are identified. As an inland site, it will be limited to launching RLVs that will not release spent rocket stages on populated areas.

South Dakota Spaceport. Western South Dakota near Ellsworth Air Force Base is being considered as a future spaceport site. However, there is no infrastructure in place and all work has been preliminary and is dependent on market development.

Southwest Regional Spaceport. The state of New Mexico proposes to construct and operate the Southwest Regional Spaceport near the town of Upham. The spaceport expects to launch vehicles capable of terminating each flight without damage and making a fully controlled soft landing under emergency conditions. No spent stages or other components would be released during normal flight.

Spaceport Washington, a public-private partnership, has identified a site in central Washington state at a former Air Force base where infrastructure is already in place. It is propose to use the site for horizontal and vertical launches.

Texas Spaceport. The state of Texas supports the development of one or more commercial spaceports for RLVs. Two candidate sites are on the Gulf of Mexico; another is in the desert of western Texas. They will be selected in cooperation with

RLV operators. The spaceports will be owned and operated by a public–private partnership. Texas established a Spaceport Authority in 1999.

Utah Spaceport. The Wah Wah Valley Interlocal Cooperation Entity proposes constructing and operating a commercial launch site that using Utah State Trust lands located southwest of Milford. The mission is to provide a cost-effective launch and recovery facility for RLVs.

Wisconsin Spaceport is being established at Sheboygan, on Lake Michigan. It began operating 4 years ago and has launched several sounding rockets. Although there has been no approval by the state for future development, the plan is to have the capability for RLV launch and recovery if the market warrants.

South America

Alcantara Launch Center, Brazil. The Alcantara Launch Center targets the expanding GEO communications satellite market. Located on the coast less than 3 degrees south of the equator, the spaceport offers a site for launches into geosynchronous orbits. The site of 30 NASA/INPE sounding rocket launches in 1994, Alcantara has had more than 275 suborbital launches since 1989, with two unsuccessful orbital attempts in 1997 and 1999. Brazil had hoped to have its first successful launch in 2002, but after failure it has been postponed for several years.

Kourou, French Guiana. The Guiana Space Center provides optimum operating conditions for commercial launches. France built its launch base at French Guiana in 1964 and became operational in 1968 with the launch of a French Veronique sounding rocket. Launching near the equator reduces the energy required for orbit plane change maneuvers. This saves fuel, enabling an increased operational lifetime for Ariane's satellite. The ELA-3 launch facility was developed for Ariane 5, with the first mission performed from this site in 1996. This facility is designed to handle 10 or more Ariane 5 flights a year.

Russia

Russia's extensive activities in space since the launch of Sputnik means that it has numerous launch sites. These sites include Baikonur, Plesetsk, Kapustin Yar, and Svobodny Cosmodromes. Plesetsk Cosmodrome, a longtime military launch site, and the new Svobodny Cosmodrome are now used as commercial launch centers. The Russian Soyuz, Proton, Start, Rokot, and Kosmos boosters are used. Russian companies have also begun to use military launch centers as commercial spaceports since the collapse of the Soviet Union. The country is aggressively marketing its ELV launch capability, particularly the Proton launch system based at the Baikonur Cosmodrome.

Kapustin Yar is located south of the city of Volgograd on the Volga River, and the launch center, founded, in 1947 is the home of Russia's first ballistic missile launch site.

Plesetsk. In 1957, the Soviet government approved the creation of secret missile bases code-named 'Angara' near Plesetsk. As of 1997, Plesetsk had launched more than 1,500 spacecraft. Because of reduced defense spending and military space activities, space launches fell from 47 in 1988 to six in 1996.

Baikonur Cosmodrome in Kazakhstan was the Soviet Union's premier launch facility and is still the site of Russia's crewed space program. Its high-latitude location in central Asia reduces orbital payload capability by as much as 35% over an equatorial location. With the collapse of the Soviet Union, the possibility of transferring crewed launches from Baikonur to Plesetsk was considered but not implemented because of the cost of transfer and the location of Plesetsk.

Asia

All of China's launch sites – Jiuquan, Taiyuan, and the Xichang Satellite Launch Centers – use the Chinese Long March launch vehicle. China has launched its own satellites since 1970. The first commercial launch was in 1990, when a Long March rocket placed a communications satellite into orbit for a Hong Kong company. The Chinese hold 6% of the world's commercial launch business and offer a discount to customers using the Long March vehicle. Trade agreements that limit the number of western satellites China can launch expired in 2001, which could give the Chinese a greater share of the world market.

The rapid macroeconomic being experienced by China, coupled with its ability to get a man into space, has led to an enhanced national pride in aerospace with the likelihood that considerable resources will become available to further develop its commercial activities in space.

Japan's launch centers are located at the Tanegashima Space Center on Tanegashima Island and Kagoshima Space Center on Kyushu Island. Japan began launching small rockets in the 1960s and developed its H-2 rocket program in the 1980s and early 1990s. The H-2A was first successfully flown in 1994.

Future Demands for Commercial Space Transportation

There are a number of possible demand drivers for space transportation. Some of these would seem to be of relatively immediate interest while others are more long term.

Energy

A number of alternative space-based energy programs have been considered – including nuclear research laboratories, solar energy stations, and nuclear energy generation stations at low-Earth or geosynchronous orbit. Using space as a location for power generation systems requires heavy-lift capability beyond that available or projected in the most ambitious forecasts. The difficulty is compounded because current robotics technology is incapable of supporting the assembly of the required space structures, thereby requiring a prohibitive amount of astronaut extravehicular activity. A possible exception would be the development of orbit-based nuclear research facilities to support the use of ground-based nuclear energy.

If necessary, research facilities could be built using existing launch capacities and the robotic technology developed for the space station. The use of space-based services to conduct energy research, production, and distribution could, however, significantly increase launch rates as early as 2020 and could require hundreds of flights a year by 2050. Launch infrastructure would have to be improved to support these massive space-based ventures.

What would drive the use of space services as an adjunct to terrestrial energy production?

First, the demand for energy continues to increase as the Earth's population grows and Third World countries become more industrialized. Total energy consumption is expected to double in the next twenty years and again in the following twenty.

Second, excluding nuclear energy, a decreasing amount of energy is available from Earth sources. Government estimates show that the world's crude oil reserves are slightly over 1,000 billion barrels, which approximates the amount consumed to date (Department of Energy, 2001). The peak discovery of global oil occurred in the 1960s; recent discoveries represent only about 25% of the oil needed to sustain current rates of usage. Earth's inhabitants are now consuming 26 billion barrels each year – a rate that is increasing. The by-products of oil and gas consumption continue to harm the atmosphere. Renewable energy sources such as solar, wind, wood, grain, wave action, and geothermal energy are not yet sufficiently developed to support the world's energy demands.

Space provides an environment to research energy production from fission and fusion sources. The benefits of doing research in a medium-Earth orbit include providing a safe environment, a readily available vacuum source, and low gravity. A research base on the surface of the Moon – which has a large source of tritium, a nuclear fuel – or a research base in high-Earth orbit could test energy technologies without harming Earth's population.

Since the 1960s, there have been numerous proposals to build a large solar collector in space to beam energy directly back to Earth. Although technically feasible, this project is costly and challenging. At the current maximum launch capacity of the space transportation system, as many as 2,000 flights would be needed to construct a large solar collector. Significant reductions in the number of launches could occur with the development of super-light materials and structures that are self-erecting or inflated on-orbit. These materials are being tested today and offer promise for large-scale space structures. Until these advances are realized and an economical heavy-lift launch vehicle is developed, the only possible activity in this time frame is a demonstration mission.

Environmental Monitoring

Over the past 20 years safeguarding the global environment is a major and growing concern, and monitoring its health from space provides perspective and data that would not otherwise be available. An environment consists of everything that is external to an organism. Nonliving environmental factors, including temperature and sunlight, are called the abiotic environment. Living, or recently living, organisms make up the biotic environment. The abiotic and biotic environments interact to make up the total environment of living and nonliving things (National Drought Policy Commission, 2000). Ultimately, the sun provides sustenance for all life on Earth.

The declining quality of our environment is evident: air pollution, acid rain, contaminated groundwater, soil erosion, deterioration of the ozone level, and extinction of some animal species. Since society increasingly has to fund remediation measures to correct damage to human health and the environment, it would be more effective to apply some of these resources to preventive measures, such as investigations of atmospheric particles that cause sickness.

Within the last 300 years industrialization has improved the quality of life in many ways. But industrialization has required the use of large inputs of biotic and abiotic materials. It has resulted in the environmental release of large quantities of pollutants such as carbon dioxide, sulfur dioxide, nitrous oxides, and ozone-producing chemicals that have caused substantial environmental changes in selected areas. As these local problems have been recognized and corrected, it has become clear that industrialization has affected the entire planet, not just local conditions. Human industrial activities have caused negative global environmental changes, especially global warming.

To analyze changes to the environment, scientists must establish baselines for environmental systems and measure the deviations form the baselines. The baseline is difficult to establish because measurement tools were not developed when industrialization began. Massive amounts of data need to be excavated from ecosystems that have been preserved for eons. Current changes are easy to measure locally because the appropriate instrumentation is now in place.

Space-based observation of water and other resources permits analysis of human influences on the environment, monitoring of crop growths, and the detection of pollution and volcanic activity, as well as disasters caused by human factors and meteorological phenomena such as hurricanes, tornadoes, floods, fire, flooding, and landslides.

Satellites can instantaneously capture and report changing weather and environmental conditions to facilitate the safer routing of aircraft. as well as maritime traffic. Satellite monitoring permits more timely and accurate forecasting and tracking of extreme weather conditions such as hurricanes and blizzards and thereby giving early warnings.

Environmental monitoring from space can also determine the composition of the atmosphere and locate and quantify anthropomorphic sources of materials that can be harmful for life. These functions require different instruments, measurements, and operations, as well as significant resources. Additional satellite measurements are needed for crop management, blight measurements, and deforestation, among other functions. To ensure the health of the biosphere, it is necessary to understand what changes occur due to variations from both natural sources and human activities. Space provides a platform from which Earth's environment can be surveyed, measured, and catalogued.

There is likely to be a significant growing demand for launches to create a band of environmental satellites orbiting the Earth at various altitudes. With instruments in every conceivable spectrum, the various aspects of the planet's health will be measured so that precious resources can be maintained and conserved. This could be accomplished with a few large satellites or many smaller, specialized satellites. Cost could be a deciding factor in which types of systems are deployed.

Global warming may be the most significant long-term environmental problem. The use of space-based observations can assist in the efforts to baseline the effects of global warming, and the medium of space can offer some solutions. The UN-sponsored Intergovernmental Panel on Climate Change (2001) finds evidence of a trend toward a warmer climate, perhaps linked to two human activities – industry and agriculture. Space has the potential to provide unlimited clean energy to inhabitants of Earth, eliminating the greenhouse gas contributions from motorized industrial sources and power plants. Elimination of agricultural greenhouse gas is harder to manage.

The most promising Earth-based solution may well be the increased use of nuclear energy to reduce carbon dioxide emissions.

The US Department of Energy's Office of Nuclear Energy, Science and Technology has conducted numerous studies over the years on various power options. Under the nuclear option with a stabilization scenario, current nuclear plants would be permitted to operate through their current licenses, license renewals allowed, and new reactors could be built if their prices are competitive. In this scenario, carbon dioxide stabilization would restrict emissions for the US to 1990 levels by the year 2010 and beyond, raising nuclear energy's share of the nation's electricity generation. Under such a scenario, space would make its contribution to earth-bound efforts by improving monitoring of key environmental indicators, power generation platforms, and developing the use of extraterrestrial sources of material to generate power.

Public Space Travel

When Dennis Tito blasted off in 2001, in a Soyuz spacecraft bound for the International Space Station, news headlines proclaimed that the era of 'space tourism' had arrived. Although Tito's trip was momentous, there is little prospect for a genuine space tourism industry in the foreseeable future. While a handful of wealthy individuals may follow Tito's lead, their space voyages will not signal a new era in space.

There are currently a dozen teams working on privately funded initiatives to build and fly experimental vehicles that will lead to commercially viable passenger spacecraft. However, no investor groups have come forward with the several billion dollars that will be required to develop a commercially viable passenger space vehicle.

This is not to say the subject of public space travel is being ignored. There are a number of organizations currently exploring ways to make it a reality. The Space Transportation Association has established a Space Travel and Tourism Division to continue the work undertaken jointly by Space Transportation Association and NASA in the *General Public Space Travel and Tourism Study* (O'Neil et al, 1968). Earlier, the Commercial Space Transportation Study Alliance of six aerospace companies sponsored a cost-oriented study of the market potential for space tourism. The study concluded:

- To achieve an economically viable payback, the current $10,000 'a pound to orbit' cost of flying payloads to low-Earth orbit must be reduced to $600 per pound or lower.
- The projected annual demand for space tourism, assuming economically viable launch costs and a ticket price of $25,000, ranges from 22,000 individuals who would sign on for a 1-day cruise on a launch vehicle to 65,000 who would sign up for a 5-day vacation.
- The commercial launch industry should eventually transition from a defense- and science-driven enterprise to be consumer-driven.

The Japanese tourist agency estimates that 5 million people a year could be launched into space by 2030 in a global industry that could be worth as much as $100 billion. A Japanese travel firm estimates package deals could eventually bring costs down to under $3,000 per passenger. Numerous studies have shown that there is considerable interest in space tourism, even at considerably higher prices. In addition, MirCorp has announced a plan to launch a private space station for paying tourists. The new space station, called Mini Station 1, will cost about $100 million and is expected to enter orbit in 2004. It would be manufactured by the Russian company RKK Energia, that built the Mir and owns a majority stake in MirCorp. The rest of the venture belongs to private investors. The space station would be slightly bigger than Russia's Soyuz space capsule and could accommodate three visitors for up to 20 days. Soyuz capsules that ferry crew members to the International Space Station could dock at the private station and then continue on to the ISS.

These types of efforts enhance cooperation and increase dialogue between government and the private sector. They also help maintain public awareness of space passenger travel as an important goal of the space program. Safety will be a critical factor in the effort to open space travel to the public. The FAA is expected to have a central role in assuring that passenger safety standards are defined and followed.

While an acceptable level of risk has yet to be determined for public space travel, it is expected that the accident record of the early years of passenger aviation will not be repeated – between 1927 and 1939 there were 119 aircraft accidents in the US claiming 255 lives. In contrast to civil aviation, where safety procedures were developed, tested, and refined as flight schedules ramped up, the space industry has 40 years of experience in human space flight and fully tested safety procedures have been integrated into all aspects of prelaunch processing, launch, and in flight and landing operations.

National Defense

Warfare in space is no longer science fiction. The US Congress and the President have recognized that command and control of space is a vital asset for national security. Military experts have called the Desert Storm campaign the 'first space war.' In Kosovo, the second space war, the Global Positioning System was used to direct precision-guided munitions to selected targets.

The 'Rumsfeld Report' (Commission to Assess United States National Security Space Management and Organization, 2001) affirmed these results, concluding that space has assumed such a major role in the nation's defense that consideration should be given to developing a separate branch of the armed forces dedicated to space.

The Report recognized that the US military is dependent on space-based assets for its survival and that it must put the proper management in place. The Report also recommended the development of a space-based antiballistic missile defense system. It concluded that it is in the US's national interest to promote the use of space to support domestic, economic, diplomatic, and national security objectives in order to protect and defend US space assets.

For the foreseeable future, the DOD will continue, for budgetary and other reasons, to use commercial space assets to fill military requirements. Commercial communications as well as weather and new promising hyperspectral imaging satellites will play an important role in national defense. Other commercial opportunities are being explored.

The list of defense-generated research programs and flight opportunities, is virtually endless. Potential missions include mesoscopic machines, kinetic weapons, space-based lasers, clusters of nano satellites, robots, power generation and storage systems, smart materials and structures, ultra-lightweight materials, new sensors, on-orbit data processing and artificial intelligence, launcher and propulsion developments, on-orbit servicing capabilities, surveillance, and both defensive and offensive technologies.

In a political climate favoring space research and exploration, DOD will undoubtedly pursue many of these topics. The list may also include the need for a rapid deployment capability for personnel, equipment, and supplies to any area of conflict in the world. This requirement is close to the commercially driven rapid package delivery capability and to the proposed faster Boeing Company commercial transport aircraft. Although much research and development can be conducted on the ground, the absolute proof of concept frequently is to be had by testing in flight or on orbit. Here there is a convergence of the common need for all space users, defense or commercial, to reduce the cost of flying a mission.

Education

Space exploration and education are traveling along parallel paths. Space programs need highly skilled workers, and educational systems need teachers who can motivate and encourage students to study science, mathematics, technology, and engineering. Space exploration provides that stimulus. Students and teachers alike can become involved in space missions, applying theoretical knowledge of mathematics, science, technology, and geography to real-life applications. Educational tools developed from space applications will prepare today's students to meet the challenges of their future.

Conclusion

A robust civil space program is imperative for the economic well-being and security of the US. The variety and volume of missions that could be flown in the coming years are large. Staying bullish on space will require determination and a concerted effort to overcome the high cost associated with space flight. There is every reason to believe that new and exciting missions, unheard of today, will emerge to take full advantage of the characteristics of space. However, unless concerted efforts are made to change the way commercial space efforts are supported and encouraged by the government, stagnation is a prospect.

The primary cause of this is the cost of space flight, which has remained stubbornly high and likely will remain so. Space travel has a long way to go, technically and operationally, to equal the accomplishments of the air transportation industry (Department of Commerce, 2001). Today's extensive commercial air transportation sector has far exceeded the expectations of its most optimistic early proponents (Heppenheimer, 1995).

The first attempt at commercial supersonic flight, Concorde, was a technological success. Now Airbus Industries plans to build a mega 680-passenger aircraft. This progress came as the result of a synergistic relationship among government, the aerospace industry, airport operators, and the public. When the traveling public of the 1930s demanded faster, safer, and more reliable aircraft, airways, and airports, the industry and the government responded. Without this collaboration and billions of taxpayer dollars, commercial air travel would be far from what it is today.

Space travel can now be equated to the air commerce of the 1920s – a high-cost, high-risk, low-volume, and novel travel experience. In the launch vehicle arena, the industry is still living off investments made in the 1950s and 1960s (International Space Business Council, 2001). During this golden era, all of today's major launch systems were developed or planned, and they or their second, third, or forth derivatives are carrying the 21st century payload to orbit.

A half-century has passed since the US last made a major investment in launch technology. It can be argued that the $1 billion invested in the now-defunct X-33 program was such an investment, but that the program was largely intended to help an ailing industry. The government's intention to invest up to $5 billion in incremental improvements to existing launch systems seems to fly in the face of the reality that revolutionary, not evolutionary, improvements are needed to significantly reduce access cost and maintain a high safety factor while stimulating new customers.

Acknowledgement

This chapter represents a shorter version of material produced by the members of the George Mason University Continuing Career Program. Beside the authors it is important to recognize the contributions of David Dickensen, Lawrence Jessie, Frank Manning, Harold Miller, John Mulcahy, James Ream, Eric Rhodes and Christine Rogers.

References

Commission to Assess United States National Security Space Management and Organization, (2001) *Report*, Department of Defense, Washington DC.

Department of Commerce (2001) *Trends in Space Commerce*, Department of Commerce, Washington DC.

Department of Energy (2001). *World Crude Oil and Natural Gas Reserves, International Energy Annual*, Department of Energy, Washington DC.

Federal Aviation Administration (2001) *The Economic Impact of Commercial Space Transportation on the US Economy*, Federal Aviation Administration, Washington DC.

Federal Aviation Administration (2001) *The Economic Impact of Commercial Space Transportation on the US Economy*, FAA, Washington DC.

Federal Aviation Administration and Commercial Space Transportation Advisory Committee (2001) *2001 Commercial Space Transportation Forecasts*, Federal Aviation Administration and Commercial Space Transportation Advisory Committee, Washington DC.

Federal Aviation Administration and Commercial Space Transportation Advisory Committee (2002) *2002 Commercial Space Transportation Forecasts*, Federal Aviation Administration and Commercial Space Transportation Advisory Committee, Washington DC.

George Mason University NASA Continuing Career Program (2002) *Spaceport Infrastructure Handbook*, George Mason University, Fairfax.

Heppenheimer, T.A. (1995) *Turbulent Skies: the History of Commercial Aviation*, John Wiley and Sons, New York.

Intergovernmental Panel on Climate Change (2001) *Report on Climate Change*, IPCC, Geneva.

International Space Business Council (2001) *2000 Space Industry Report*, International Space Business Council, Arlington.

National Drought Policy Commission (2000) *National Drought Policy Commission Report, Government Printing Office*, Washington DC.

O'Neil, D., Bekey, I., Mankins, J.C., Rogers, T.F. and Stallmer, E.W. (1998) *General Public Space Travel and Tourism Study*, NASA, Washington DC.

Vision Spaceport Partnership (2000) *Spaceport Concept and Technology Road Mapping*, National Aeronautics and Space Administration, Cape Canaveral.

Chapter 8

VentureStar

Clay R. Hicks

Introduction

Space travel is extremely expensive – exceeding $20,000 per kilogram of payload carried. Using a multi-stage rocket to launch into orbit means large sums of money are lost. A single-stage-to-orbit (SSTO), spacecraft is logically one way of overcoming the problem. With modern engine designs (a linear aerospike engine built by Boeing with a radical new chamber-less design was to be used that would be much lighter than the Space Shuttle Main Engine and could be integrated more efficiently within the lighting body of the vehicle) and light, but strong, materials being manufactured, such a concept was thought possible. It was not, however, entirely a novel idea. Eugen Sanger proposed an aircraft to be boosted into orbit by rockets and gliding back to land on Earth in 1944.

The projected market for commercial space activity in the first decade of the 21st century has shrunk considerably in the last several years making efficiency in launch even more important. However, there probably would not be sufficient demand to warrant the development of reusable launch vehicles for at least another 10 years. However, it is worthwhile to evaluate what impact reusable launch vehicles may have by examining the following case study of the VentureStar project, which was cancelled in March 2001 after several years of effort.

It is the spacecraft designer's aim to build a space ship that is economical to launch and that can be used again and again like an airliner. For 20 years, the National Aeronautics and Space Administration (NASA) has relied on the Space Shuttle, a craft that is half aircraft and half rocket. The winged orbiter section can return from space and land on a runway, but it is costly to operate. To reach orbit it needs a pair of solid rocket boosters that are jettisoned, along with a huge external fuel tank that can be used only once. The boosters have to be retrieved from the ocean, and both the boosters and the orbiter require lengthy maintenance between flights. The Space Shuttle, although an advanced piece of engineering has failed to meet economic criteria as each launch costs $500 million, involves thousands of people, and takes 3 months to prepare.

In the 1990s, NASA pursued the design and development of a Next Generation Launch Vehicle (NGLV) that would be easier to service and cheaper to operate. In July 1996, the Lockheed Martin VentureStar Corporation was awarded a NASA contract to build a prototype (X-33) of a new spacecraft called VentureStar, that would be designed to replace the Space Shuttle when developed and proven capable (Cast and Amatore, 1996). It was intended that the prototype would only fly 15 times, lifting off from Edwards Air Base in California and landing in Utah and Montana. While the initial funding would be a public-private partnership (Lockheed Martin contributing 20% of the cost) the prototype was intended to form the basis for Lockheed Martin to develop the full sized operational version of the vehicle (Figure 8.1 gives an image of the original design of the X-33).

(*Source*: NASA History Office)

Figure 8.1. Image of the VentureStar Vehicle

The wedge-shaped VentureStar would be designed to take off vertically like a rocket, go into orbit using only one set of engines, and fly back to the ground and land horizontally like an airplane (*Spaceport News*, 1998). It could be used for manned or unmanned missions.

The VentureStar would be a fully reusable SSTO vehicle that would not need to jettison any components at any stage of flight. This feature alone would save money. In addition, only a small maintenance and ground control team would be needed to service the ship. The estimated costs of reaching space and returning would be reduced by a factor of 10.

Its design is based on experimental lifting body gliders that can fly without wings. The fat, rounded body shape – known as a lifting body – generates its own aerodynamic lift so that the ship can fly safely to a touchdown at the end of its mission. VentureStar's ultra-light construction helps make the wingless shape possible.

On reentry, the combination of high aerodynamics lift and lightness would allow the vehicle to glide through the thin air at a high altitude, taking a shallow reentry path. The entry heating would last longer, but would be less severe than was the case with the Shuttle, so less of the airframe would need protection. The VentureStar would not require humans to steer it. Onboard computers would handle every aspect of a mission, including launch, reentry, glide phase, and final touchdown on an airfield runway.

In 2000, the half-size suborbital test version of VentureStar, X-33, was to begin a series of technology demonstration missions to prove that the VentureStar concept could work. However, in March 2001, NASA cancelled the X-33 flight test program because of technical development problems and funding constraints.

The promise of an economical launch capability and everyday operations of spacecraft like a commercial airline still remains a dream. However, the following sections will discuss the development of the Reusable Launch Vehicle (RLV) as the desired goal for the NGLV and the operational issues that must be incorporated into the design of the RLV to accomplish the low-cost and quick-turnaround ground operations required for the NGLV. In addition, an overview of the proposed VentureStar Spaceport Operations concept is presented as a viable example of a spaceport-type site and facility plan that could be expected in the 2000–2020 time period.

The development of the prototype and planning for the full operational VentureStar did involve a significant number of interacting considerations.

Operational Issues

During the 1990s, NASA, the US Department of Defense (DoD), and the US aerospace industry responded to the National Space Transportation Policy's stated goal of providing reliable and affordable access to space in the development of future space transportation vehicles. In this framework, NASA was given the role as 'the lead agency in the development and demonstration for the next generation reusable space transportation systems, such as the SSTO concept' (Presidential Directive, 1994).

In the development of the RLV concept, a joint government/industry Operations Synergy Team (OST) was commissioned to ensure that lessons learned from previous space transportation programs and studies would be applied to the NGLV. The OST identified certain operational issues that must be dealt with in order to achieve the SSTO goals of reliable low-cost space transportation and order-of-magnitude reductions in operating costs. From an operational perspective, according to the OST, this goal can be achieved only when system operational functions drive vehicle design.

The OST then developed an RLV Operations Concept, 'Vision,' comprising ten goals focused on operational improvements geared toward minimizing ground test time, resource dependence, and servicing requirements while maximizing vehicle self-diagnosis and dependability. The research and technology development of the RLV would be focused on those areas that minimize the number of subsystems and fluids used, the total parts count, and the amount of testing required to validate system integrity during ground turnaround operations (National Aeronautics and Space Administration, 1994).

The 10 goals identified by the OST for launch vehicle design that would enhance the ability of a future RLV to meet operational performance objectives, while maximizing operational utility and minimizing operational costs, are (National Aeronautical and Space Administration, 1994):

- Provide a simplified, very highly automated vehicle that would enable minimum periodic and repetitive maintenance (aircraft type) and a resultant short turnaround time between missions (hours, not months).
- Strive to isolate vehicle ground processing from dependence on facilities and ground support equipment (GSE). Routine scheduled turnaround should be to replenish consumables only.
- Promote vehicle health monitoring/management systems and self-test at a level that supplies only operations and maintenance anomaly information that requires corrective action prior to the next flight. Let the vehicle 'talk' to the ground remotely for processing and maintenance needs.
- Eliminate flight readiness-style vehicle certification for every flight. Provide aircraft-style vehicle type certificate for repetitive commercial flight operations.
- Strive to eliminate unplanned work by designing in performance margins and flight hardware allowances to eliminate processing impacts. No dedicated software maintenance function is required to support operations.
- Reduce operations and hardware complexity for maximum utilization of ground resources and decrease the opportunity for human-induced system failures: less 'hands-on,' less human factor.
- Employ nearly autonomous ground management planning at top levels. Focus on automatic interactive scheduling of flight vehicles, ground support facilitators, and support logistics.
- Adopt minimum standardized payload interfaces to assure maximum flexibility and affordability. The most affordable vehicle would be blind to payload needs like a truck. Decouple payload processing from vehicle processing to eliminate payload impact on the launch vehicle infrastructure.
- Ensure joint participation and application of the synergism available among operations, avionics, propulsion, payloads, and vehicle design to the preliminary architecture/vehicle concept and the operations development process – i.e., Integrated Product Team–IPT. Facilitate the identification and sharing of technologies that can enable development of a vehicle system to meet the attributes of the National Space Transportation Policy.

- Assign the role of engineering (concept, development, and technology) to be responsible for continuous improvement and technology advancement for future market-driven needs.

It was hoped that the goal of driving the total space transportation system design, rather than reacting just to the vehicle design, would result in the realization of the economic goals of providing reliable and affordable access associated with the operation of future reusable space transportation systems.

Reusable Launch Vehicle Test Demonstration Programs

The pursuit of the NGLV system needs to show that any new system really can be maintained by a small launch crew and turned around quickly to execute repetitive launch missions similar to an aircraft.

In 1991, the Strategic Defense Initiative Organization, subsequently to become the Ballistic Missile Defense Organization, initiated a flight test project to assess the operability of a SSTO demonstration vehicle. The Strategic Defense Initiative planners felt that the nation needed a less costly, and more reliable and responsive space launch capability than that offered by the existing expendable launch systems. McDonnell Douglas was selected to implement the project based on its efforts for the development of a wingless, conical SSTO vehicle that could take off and land vertically using rocket engines.

The Delta Clipper–Experimental (DC-X) program initiated on a $60 million budget, developed the DC-X vehicle that was a one-third-scale flying model of a vertical takeoff/vertical landing SSTO. It took off and landed in 11 tests, during which it never reached supersonic speeds, but it did successfully demonstrate the attitude reversal maneuvers needed to land.

The DC-X's biggest success was that it flew twice in one day with its entire ground support team consisting of only 12 people. Even though the DC-X was not a full-scale SSTO, the promise of a quick turnaround vehicle with a significantly smaller ground support team was a step closer to being realized.

The DC-X program suffered a test failure that damaged the vehicle, and no additional funding was approved for it. However, in 1994, NASA took over the DC-X program and decided to pursue the SSTO as the NGLV, initiating its own flight test program called X-33. In order to take a bigger step than DC-X, the X-33 would be a two-thirds-scale program to fly into space and reenter the atmosphere at Mach 15 while still operating at suborbital velocities.

In 1996, the Lockheed Martin VentureStar Corp was selected by NASA to build and fly the X-33 flight test vehicle. The Lockheed Martin concept was the most technologically challenging, having a unique lifting body shape that combined wing and fuselage into a single structure, an innovative metal thermal protection system, and a new unknown type of rocket engine called an aerospike. The X-33 would take off vertically and land horizontally, much like an airplane (Figure 8.2). Lockheed Martin announced that if awarded the X-33 contract, it was prepared to finance the X-33 program by spending $4.5 billion of the company's own money to develop it into a full-scale VentureStar launch system capable of SSTO flight (Zurbib, 1999).

VentureStar Spaceport Site Qualifications

In addition to the need for the vehicle itself, there would be the need to develop adequate and appropriate land-side facilities to handle preparation, take-off and landings. New spaceport facilities were thus required.

The term spaceport encompasses facilities directly related and essential to servicing spacecraft, enabling spacecraft to take off or land, and transferring passengers or space cargo to and from spacecraft, but only if the facilities are located at, in close proximity to, or in the direct logistical support path of the launch or landing site to perform these functions. The term also includes other functionally related and subordinate facilities, such as launch control centers, repair shops, maintenance or overhaul facilities, and spacecraft assembly and storage facilities that must be located at, adjacent to, or in the direct logistical support path of the launch or landing site.

(*Source*: NASA History Office)

Figure 8.2. Image of VentureStar Releasing a Satellite

In July 1998, the Lockheed Martin VentureStar Corporation conducted a briefing on its VentureStar Spaceport Concept to a meeting of over 30 different site representatives from the US.

The objective of the meeting was to present the VentureStar Operations Concept requirements to the candidate spaceport sites in order to acquire information on the various sites and to assess the implications of those sites on system performance (vehicle and payloads), cost and revenue, schedule, and risk. Spaceport site qualifications assessments, to be conducted and submitted to the VentureStar team, were to include, and were not limited to, the following: site description, VentureStar accommodations and enhancements, economics, schedule, and regulatory compliance.

Site Description

The assessments were to describe the spaceport sites' geological characteristics, taking into account soil types, soil stability, and subsurface chemistry, the seismic history and projections, and expected tidal changes. Environmental concerns, such as known existing contamination and past work; endangered species, historical sites, and wetlands; and noise abatement and air quality issues were also to be addressed.

VentureStar Accommodations and Enhancements

The proposed conceptual site plan would map the general arrangement and note the use of the planned improvements and the need for potential additional improvements. Vehicle performance-related site features would include latitude, longitude, and elevation of major site facilities. Landing approach corridors were to be assessed as to obstructions, lines of sight, communications, and other factors. Shared facilities would be utilized and identified when they offered distinct advantages.

Public safety concerns meant that there would be detail population data for over-flight areas outward from the site. Climate characteristics, including past history, would have to be assessed as to their impact on vehicle and payload processing and movements between facilities. Local, regional, and state support resources should be applied to the site (e.g., utilities, cryogenics and gases, and community services).

Financial Factors

Financial efficiency requires appropriate appraisal techniques be applied, treating the investment in a genuine long-term context. An analysis of the cash flow over time including taxes and fees, utility rates, and incentives and other investments must be conducted. This is necessary to assess the payback period, break-even point, and any establishment of a rate of positive cash flow in order for the spaceport site operations to be an economically sound business.

Schedule Support and Regulatory Compliance

This section of the proposed spaceport site assessment would address site commitment, environmental conditions, site geology, and existing improvement implications. The regulatory compliance section would define the agency requirements, establish the processes involved, and detail the agency's capacities to negotiate, grant, and enforce compliance factors.

The VentureStar team emphasized that VentureStar would be a new revolutionary space transportation concept and, as such, would require a new way of doing business. The siting of any initial VentureStar operations complex would depend heavily on systems performance enhancements, greatest attained market share (which means a site must accommodate the broadest range of addressable markets, including International Space Station, GTO/LEO, medium inclination, and high inclination, and the best financial and business plan).

VentureStar System Operations

The initial VentureStar fleet would consist of two vehicles (see Chapter 7 for details). The operational capability goal of the system would be forty flights per year. Initial operations would support 20 flights per year for the two vehicles. A turnaround time of 7 days would be the planned processing time from landing to launch, with a 2-day turnaround capability for special customer requirements as needed.

Initially, two spaceports would be required for the two vehicles, and these would be located next to each other. Depot maintenance would be needed for each vehicle after 20 flights. Payload processing would be de-coupled from the vehicles, and the payload would be mated to the vehicle through the use of a special Payload Mission Module (PMM). The PMM would be externally mounted to the vehicle and installed on the vehicle during the final hour of the 6- to 9-hour preparation time prior to launch.

Fewer than fifty people were required for vehicle and payload touch labor during processing, with a total complement of about 300 people involved in the two vehicles and spaceport operations activities. While this number is not large, it, with the associated halo effects, would be enough to stimulate competition between potential spaceport locations in at least 25 states in the south and west of the US.

Operations Approach

The VentureStar spaceport's operation is seen as very different to the current facilities in use. The vehicles would take off from the spaceport where they land. Each spaceport would conduct onsite processing of payload elements into a PMM. The spaceport would also produce the vehicle consumables onsite. A vehicle onsite final assembly building would be utilized at all spaceport locations (build where you fly).

The entire VentureStar system (flight vehicle and spaceport) would maximize the use of automation for built-in system operability of the vehicle and spaceport in order to enhance processing and minimize human involvement. The major business goal of the VentureStar Corporation was to maximize the collocation of all the VentureStar activities.

Onsite Activities

There are a significant number of major onsite activities that would be required at the spaceport. These include:

- Vehicle final assembly;
- Vehicle processing and maintenance;
- Payload assemblies processing;
- Flight planning and control operations;
- Spaceport operations and maintenance;
- Support systems operations and maintenance;
- Business administration and management;
- Marketing and sales;
- Testing and training.

Some of these requirements can be looked at in more detail.

Vehicle Final Assembly. The very large size of the vehicle does not allow delivery of a fully assembled vehicle. Some on-site assembly would be required. The total onsite activities for final assembly would be determined by the system, subsystem, and component size, economic trade studies, and the sequence of final assembly activities.

Vehicle Processing and Maintenance. The operations flow for each vehicle mission would include the following major onsite activities:

- Horizontal landing and vehicle safeing;
- Transport to the mobile translating shelter and attachment to the launch; mount strongback system;
- Removal of mission module;
- Next mission module installation;
- Rotation to vertical takeoff position;
- Loading of consumables and mission software;
- Roll back mobile shelter;
- Takeoffs;
- In-flight operations;
- De-orbit and reentry;
- Depot maintenance – conduct in-depth inspections and refurbishment as required and install upgrades.

Payload Assembly Processing. For full operational and economic efficiency the division of operational activity between the spaceport and the customers would need to be fairly clear and straightforward.

The customers of the facility would plan, operate, maintain, and supply logistic support for their own equipment. They would have contact with the spaceport to schedule support services and would provide the final approval for launch of the payload.

The role of the spaceport would be to plan, operate, maintain, and provide the logistic support for the launch facilities as necessary for the conduct of an efficient and successful flight. Flight and ground safety as well as launch licensing and environmental compliance would be spaceport responsibilities.

The key activities to be performed by the customer would include:

- Process, return mission module. Remove and deliver returned payloads, and refurbish mission module.
- Process next payload and mission module. Prepare mission module; prepare spacecraft, ASE, and upper stages; conduct integration and encapsulation of all payloads.
- Conduct fueling and hazardous materials processing.
- Conduct delivery of mission modules for insertion into vehicle.
- Conduct ISS crewmember flight preparation and post flight processing.
- Conduct biological payloads preparation and post flight processing.
- Conduct last-minute boarding of ISS crew and time-critical payloads.

Support Systems Operations and Maintenance

The support systems operations and maintenance functions would include: operations and maintenance of systems for the production, storage, distribution of fuels and oxidizers, pressurant and purge gases, and electrical energy J-7; receipt storage and distribution of spares; receipt and dispose of waste materials; and communications operations and maintenance for all ground systems and vehicle tracking and telemetry systems.

VentureStar Spaceport Description

VentureStar would require spaceport facilities that are somewhat different to those of the Space Shuttle. It was seen not least as a vehicle that could be very quickly turned around after a mission. It was not to be an infrequent voyager of the kind that the Shuttle is.

Spaceport Requirements

The spaceport operator would be responsible for safety, capacity to accomplish the forecasted mission throughput, assurance of spaceport sites commonality in critical components, maintenance of business plan support functions, and planning for growth accommodations.

Spaceport Approach

The spaceport operator would develop minimal facilities, systems, and equipment to ensure multiple uses and reduced maintenance costs. The spaceport operator would also develop and maintain a small operations staff that would be multi-skilled to maximize productivity across all spaceport activities and would foster a bias for automation. The operator would also have a coordinating role in combining tasks and activities into appropriate groupings to reduce overlap and redundant operations. The spaceport operational team would need to be able to maintain a robust and high propensity for sustained activities under considerable schedule demands.

Site and Infrastructure

The spaceport site and infrastructure needs and requirements were sized by the VentureStar team during its construction of the spaceport site at Edwards Air Force Base, California, for the X-33 flight demonstration tests, which were cancelled in March 2001. The total site area would require some 5,000 acres of land. About 3,000 acres would be utilized for an arrangement to accommodate two launch pads with a clear zone, typical for each pad during launch, consisting of a 15,000-foot radius for siting the support facilities.

Access to the spaceport site would entail a number of capabilities and specifications:

- Roads (AASHTO HS20 loading);
- Rail (standard main line, Cooper E80 loading);
- Barge (TBD – no current requirement);
- Air (runway capable of supporting typical transport aircraft);
- Utilities – electric power (41 MW from redundant sources, potable water (100,000 gallons per day), process water (400 gpm average flow), fire protection (3,400 gpm peak flow), natural gas (3,400 cfm), waste disposal (14,000 gallons per day), and communications systems (TBD);
- Community Services – fire protection, law enforcement, health services, and emergency response.

General Arrangement

The defining characteristic of the layout of a VentureStar spaceport is the requirement for a clear zone of a 15,000-foot radius from each pad during launch. Assuming a square area with 15,000-foot sides with each of the two launch pads in a bottom corner of the square, and a 10,000-feet by 200-feet runway just above the intersection of two circles drawn by the clear zone of 15,000-feet radius from each launch pad, one has a general idea of the spaceport arrangement.

A perpendicular access road from the runway runs down to a horizontal road that connects the two launch complexes (15,000-feet apart) at the bottom of the imagined square. These are the major pathways that connect the spaceport facilities and facilitate the processing and servicing of the two vehicles.

The facilities that are common to the use of both launch complexes are located in the area above the runway at the top of the imagined 15,000-feet-square area. Assume you are looking at a picture of the area and above the runway would be the location of the spaceport common use facilities and buildings that were previously discussed:

- The system development complex is located to the left of the area above the runway. This complex contains the final assembly facility and support buildings for subassembly, subsystem testing, composites manufacturing, and painting, and offices for the vehicle final assembly refurbishment and maintenance activities.
- The center area just above the runway is the site of the operations control center for the vehicle, the control tower for the runway, and an aviation facility that supports air transport operations. The spaceport administration building and the site maintenance facility are also located in this center area, above the operation center, the control tower and the support facility.

  Additionally, the logistics and support facility are located in the central area, just below the runway. This facility supplies transportation services to the flight vehicles as they are moved from the runway to the required processing facility and the two launch complexes for final takeoff.
- The payload processing facility and the hazardous payload processing building are located in the right area above the runway. The hazardous payload processing building is set away from other facilities for safety.

Launch Complex

Preparing for a launch is a multi-stage procedure. The two launch complex areas, identical in design and layout, would each be centered on a 240-feet by 1,300-feet concrete flat surface. This surface supports the mobile translating shelter located at one end and the launch pad at the other. They are connected by a translating shelter rail system that would move the RLV, which would be attached to a strongback within the shelter down to the launch pad to meet the rotating launch mechanism (RLM).

Once the RLV and strongback are attached to the RLM, they are ready for rotation to the vertical launch position. After the RLV is ready for launch, the translating shelter is moved back away from the launch pad. A vertical access tower allows final vehicle and payload activities to be completed. On one side of the launch complex are the oxygen storage and nitrogen storage systems and on the opposite side are the hydrogen storage and helium storage systems for final RLV fuel loading. The RLV is now ready for launch.

Operations Control Center and Aviation

The operations control center consists of two working levels. The first level contains the simulation laboratory, offices and work areas and the second contains a large control room where both vehicles can be supported for all flight phases at the same time. A training and mission planning room is also located on the second level, and a viewing gallery is situated at the front of the building where there are observation windows. The control tower and aviation facility are in this vicinity and support the air transport activities for the spaceport.

The Cancellation of the Project

The VentureStar Project was cancelled on March 31 2001. It had been an ambitious program with a tight schedule – it was meant to come into service in 2004. But it had met with a series of major technical problems that had pushed up its costs, and ultimately put into question whether its main objectives would be met. The US government had contributed $1.3 billion to the project by this time ($912 million from NASA's budget) and Lockheed Martin some $357 million. Put simply by the director of NASA's Marshall Space Flight Center, Art Stephenson, 'The cost to fly the X-33 or X-34 exceeds the benefits that could be derived from flight demonstrations of the vehicle.' (X-34 was a prototype for a much smaller reusable rocket.)

The cancellation came at a time when NASA's position was under review and during a period when the agency had experienced several years of funding scrutiny. In 1998, NASA had a budget cut for the fiscal year of $173 million, the sixth in as many years (Laurent, 1998). The agency also had a history of serious cost overruns – for example, in 1997 the US General Accounting Office reported significant cost and schedule overruns by the international space station main contractor, Boeing. As a result it had been under pressure from the US Congress to further commercialize its activities.

The situation by the time of the cancellation had worsened and, amongst other things, there was a threatened overspend of $4 billion on the $25 billion cap that had been placed on international space station expenditure. It also coincided with managerial changes at the agency. The funding for the VentureStar project had initially in 1996 been set at $912.4 million from NASA and $211.6 million from Lockheed-Martin but that had demonstrably been exceeded.

It is difficult to assess the extent to which the final outcome was the result of poor management, the degree to which the original budget was inadequate given the significant challenges in constructing the model and carrying through test launches in the time scheduled or a combination of these and other factors.

Technical issues and resulting design changes were a problem. Early changes had involved the need to redesign the vehicle to meet fuel requirements and retain space for a viable pay-load (to be 56,000 lbs into low space orbit or 18,000 lbs into geosynchronous transfer orbit). This in turn meant the need for an external payload bay and the re-designed vehicle in 1999 had more drag as a result. VentureStar's

Vice President claimed this would not delay the program when the new design emerged.

The later technical problems that were encountered by the VentureStar project were mainly focused on keeping its weight down. This was required because of the single stage design. In particular it was important to keep the weight of its fuel tanks down – given that 90% of the vehicle's mass would be fuel. The aim was to design the internal tanks of VentureStar for greater volumetric efficiency that reduced the need for heavy and wasteful supporting structures.

The US General Accounting Office (1999a) felt, even prior to a later major fuel tank problem, that 'Delays in the X-33 Program may affect NASA's investment plan for future space-launch programs, including decisions on whether and when to update the space shuttle fleet or rely on a new launch vehicles, such as the VentureStar RLV.' Later, the General Accounting Office (1999b) gave evidence to congress that it was unclear how the program would achieve the cost reductions that were then needed to bring it in within budget.

The official NASA response was that the delays in carrying out test flights – from March 1999 to July 2000 – would not affect the overall schedule for the X-33 program. This was despite that fact that results from NASA's Space Transportation Architecture Studies suggested that industry would not be ready to start production of the full size vehicle until at least 2005 – a year after the initial planned introduction into service.

The two 4,600 lbs, 29,000 gallons fuel tanks on the X-33 were composed of new forms of graphite epoxy composites but one cracked under testing in November 1999. Further analysis indicated that the tanks corroded faster than expected due to reaction with the liquid hydrogen fuel. There were deemed to be significant manufacturing flaws that were too expensive to remedy. The result was that heavier aluminum tanks were to be used. This, in combinations with other earlier modifications that had reduced the goals for the vehicle (e.g., the projected top-speed was reduced by 10% – to Mach 13.8 – and some technology demonstrations were eliminated), pushed up costs of operating the vehicle.

But there are also more fundamental issues that were important in the failure of the X-33 program. It was a essentially a single option program (the Chairman of the US House space sub-committee expressed her view of the situation in August, 1999 as 'The Administration put all its technology eggs in one fragile basket and then told it to start flying').

The General Accounting Office (1999a) was rather more detailed in its critique. It argued that the program had ignored the experiences of the past where the DOD had successfully developed the DC-X and had given the work over to NASA. Basically, the early successful phases of the X-program had involved early and frequent flights of multiple copies of competing designs that incrementally expanded the knowledge envelope. As a result the program was under-funded with only a single copy of one experimental design being constructed to confront the major technology challenges that go with the development of a single-stage-to orbit RLV. Further, economics suggests that while technology can reduce the costs, it is competition that reduces price for final consumers.

The change in strategy also came at a time when developments in commercial private sector involvement were in their embryonic stage. The cost sharing arrangements used fitted technically within the 1977 Federal Grant and Cooperative Agreements Act but NASA's use of this until the X-33 initiative had been with small research projects generally tied to universities. The agreement with Lockheed Martin limited NASA's financial exposure but with a slower growth in launch demand than anticipated meant that Lockheed Martin was unwilling to take more than the initial commercial risk. The structure became a *de facto* fixed price contract for an R&D program rather than a joint commercial venture.

Conclusions

The spacecraft designers' dream of being able to design and build a reusable launch vehicle that can safely, reliably, and economically fly crew and cargo payloads into space remains a goal that requires significant national and private funding commitments that, at the current time, appear not to be forthcoming.

However, the Lockheed Martin concept, and plans for the VentureStar program offer a glimpse of an operational space craft and spaceport concept that could significantly reduce the cost of preparing and launching payloads into space. This concept presents a new way of doing business that would involve the joint efforts of federal, state and local governments, as well as the private sector to develop the next generation of launch vehicles and the first generation of a spaceport devoted to serving the customer with affordable access to space. The problem seems to be that the sudden transition in the role of NASA away from an essentially research institution combined with new ways of financing were too much to handle at the same time.

References

Cast, J. and Amatore, D. (1996) 'Lockheed Martin Selected to build X-33,' *NASA News Release*, July 2.

General Accounting Office (1999a) *Space Transportation: Status of the X-33 Reusable Launch Vehicle Program*, GAO, Washington, DC.

General Accounting Office (1999b) *Space Transportation: Progress of the X-33 Reusable Launch Vehicle Program. Testimony to the US Congress*, Washington, DC.

General Accounting Office (2001) *NASA's X-33 and X-34 Programs*, GAO-01-1041R, GAO, Washington DC.

International Space Business Council (2001) *2000 Space Industry Report*, ISBC, Arlington.

Laurent, A. (1998) 'Proud to be Penney Pinchers,' *Government Executive. Communication*, August 1.

National Aeronautical and Space Administration (1994) *Operations Concept Vision and Operational Criteria Documents*, National Aeronautical and Space Administration/ Industry Operations Synergy Team, Washington DC.

Presidential Directive (1994) *National Space Transportation Policy*, Presidential Directive/NSTL4 Washington, DC.
Spaceport News (1998) 'VentureStar: A Bright Look at the Future,' 37 (May), pp.1-8.
Zurbib, R. (1999) *Entering Space: Creating a Spacefaring Civilization*, Pitman, New York.

Chapter 9

The FAA and Microwave Landing Systems

Kingsley Haynes and Roger Stough

Introduction

Organizational structure is important, and its importance is growing as the world has become more complex. There is a continual need for companies and official agencies to review their institutional structure if they are to be efficient suppliers of services. To do this effectively an organization needs to be able to learn, and to be able to be react appropriately to what it has learned.

In the 1980s and early 1990s over $17 billion dollars was spent by the US Federal Aviation administration (FAA) to modernize the US air traffic control system but without any increase in productivity and with demonstrated failures in the ability to transfer from old systems to newer technology. The problem goes much deeper than simply the number of air traffic controllers or inspectors, or a minor adjustment in its mission. The matter addressed here is whether the FAA had the characteristics to learn and to improve or acquire the necessary core competencies.

Few, if any, other government agencies interface more with the general public than the FAA. It is therefore especially open to public scrutiny. It is one of the few organizations that provide a physical product and a definable service that most citizens can utilize on a daily basis. Its main product is air space and the management of the airplanes that fly in it, including ensuring the safety and reliability of the aircraft and its components. Other government agencies, like the Department of Defense (DOD), have been able to procure and manage advanced technology more effectively.

While governmental bureaucracy is certainly problematic, this does not explain all of the problems that have occurred at FAA over the past two decades. The question is rather whether the problems are the product of a culture that does not recognize clients other than itself. And whether this fostered a system in which the normal checks and balances did apply, so there was accountability. Furthermore, were there effective mechanisms in place whereby the FAA could learn from its successes and mistakes, and from this make substantive changes?

The Idea of Learning Capacity

The learning capability of an organization is in its ability to create, acquire and transform knowledge, enabling it to upgrade its skills, expertise, and competencies to fulfill its objectives in a fast changing and turbulent environment. The concept can be applied to a government agency as much as to a commercial undertaking.

As organized, the FAA research and acquisition organization ran under the assumption that air traffic controllers were its clients. While these individuals are vital to the system, they are not the end-users – these are the flying public and the airlines. This is why the Administration effectively had no client other than itself.

An undertaking that serves only itself cannot learn. The problem becomes one of signals, interaction and information. If an organization does not receive the proper input from outside, no learning will occur. It is this interaction that creates the learning process, a fundamental necessity for the FAA. It needs the ability to adapt to rapidly changing conditions. Transactional learning is the missing ingredient at the FAA. The Administration is a prime example of a Fordist firm in that it concentrates almost totally on production. In short, evaluation is mostly focused on inputs rather than outputs. As long as this is occurring, those providing oversight see no problems. Since oversight has been ignored, the FAA must be able to learn flexibility in order to do its job.

The fact that FAA systems and projects were being developed on average over 5 to 9 years behind schedule is evidence of the lack of the FAA's ability to learn. If projects were this mismanaged in any other context, those in charge would have been dismissed. Signals would have been sent and changes made to assume supply-demand articulation. In the case of the FAA, it was developing a new air traffic management system where there was no accountability to those using the system. Part of the problem, however, also lay with the airlines that should have been more aggressive in working with the Administration to solve deficiencies.

Why Can't the FAA Learn?

After the May 1996 crash of the ValuJet plane in Florida, media attention focused on the FAA. Don Phillips in the June 10 issue of *The Washington Post* asked important questions in debates over the proper role of the FAA and how it should be organized. Why can the FAA not learn from previous mistakes? Why can it not adjust to the rapidly changing domestic airline industry? Everyone was aware that the FAA had considerable difficulties in its management and procurement processes. Stories about outages at air traffic control centers, overworked air traffic controllers and museum-quality computer equipment were widespread. At the same time the Administration had a capable and hardworking staff. The answers to FAA's problems went much deeper than an organizational chart or procurement hassles.

Learning Capacity

With fundamental changes occurring in industrial organizations, learning capability has become the core competency needed for competitiveness. In the context of non-commercially driven government agencies, competitiveness is the ability to manage development and meet relevant objectives more effectively than other organizational structures. The inability of the FAA to do this in the late 1980s and early 1990s lay at the heart of its problems.

The organization is only one institutional form that promotes the development and maintenance of learning capability in an undertaking. Other forms of institutional arrangement such as networks, also promote the learning capability of organizations. Michael Porter's (1990) diamond model of industrial structure for assessing competitive advantages of nations (interactions among factor conditions, demand conditions, supporting and related industries, organization structure and rivalries) can also be re-interpreted as a model of transactional structure revealing important factors that influence the learning capability of related organizations.

Changing Nature of Competition

'Fordism' was the predominant management style for fifty years from the 1920s. At the height of Fordism, competitive success depended on large-scale production of standardized products providing a cost advantage over competitors. Firms were regarded as systems that transformed inputs into outputs through pre-designed procedures and techniques. Economies of scale, special purpose machines, assembly lines and the standardization of products and processes characterized mass production methods. Business operations focused on information processing in the coordination and monitoring of production rather than on learning and knowledge-creation.

An important parallel to the division of labor was one between conception and execution. In a Fordist company, responsibilities were divided among key players within the firm: executives made business decisions aimed at maximizing profit; engineers developed and designed products and processes; and workers produced the designed products by using the techniques developed and designed procedures. Consistent with this standardization was a division of the labor force. Jobs were designed to require limited skill and knowledge. Huge pools of semi-skilled workers could be employed with minimal need for on-the-job training and replaced with minimal loss of learning capability.

Today, competitive advantage rests more on the superior capability of undertakings to improve products and processes than on any advantage in the production and distribution of standardized goods. The strategic definition of business firms is that of a learning organization or knowledge-creating company. Indeed, any factors that promote learning capability become sources of competitive advantage. A prime example in the airline industry is Southwest Airlines which achieved a superior cost advantage through productivity increases (e.g. by faster turn round of aircraft and use of secondary airports) while at the same time paying its employees competitive wages.

Ways of Learning

Fordism

In the Fordist era, especially in American mass production firms, learning was generally pursued by four dominant agents. These were: a national education and research system in charge of the public production and reproduction of generic knowledge; R&D and engineering departments in charge of technological innovation and learning within organizations; top management in charge of entrepreneurial activities and learning from the marketplace, and a mass production and distribution system in charge of exploiting economies of scale and economies of learning.

Despite its competitive advantage in radical innovation and economies of learning and scale, the Fordist organization is handicapped in some forms of learning. The focus of the system on standardization and maintenance under-exploits the opportunities for shop-floor improvement and learning. Strict division of labor and its corresponding rigid departmental structure block the integration of knowledge and skills that are crucial for effective learning and knowledge creation. Separation of conception from execution cuts away the necessary feedback from operations sites to design and R&D engineers for perfecting products and processes, impeding the mutually reinforcing process of learning. Separation of design and operation from marketing also blocks the consumer's feedback about potential improvement opportunities. And focus on information processing rather than knowledge creation diverts the manager's attention away from learning.

Further, the semiskilled nature of operations jobs, together with the adversarial nature of management-employee relationships, decreases employers' incentive to provide on-the-job training. This lessens the ability and motivation of employees to acquire firm-specific knowledge and skills. The wage rate structure, based on time or piece, rather than the improvements achieved, further deteriorates employee motivation to learn. Finally, with an addiction to the principles of competition and a lack of trust and cooperation, the Fordist firm maintains a short-term contractual relationship with its suppliers. This arrangement hinders effective learning along the chain of production.

The Fordist's focus on control makes things worse. Senge (1991) points out, 'unfortunately, the primary institutions of our society are oriented toward controlling rather than learning, rewarding individuals for performing for others rather than for cultivating their natural curiosity and impulse to learn.'

It is because of this addiction to the mental model of division and control, that discovery of the learning phenomenon was theoretically constructed as a curve in the Fordist age. It was passively related to the accumulation of quantities of output, while the focus of learning was narrowed to decreasing unit costs.

The learning involved in the 'learning curve' is only a pseudo-form – passive and adaptive. The more important forms, are those that are active, systemic, integrative and generative. Rather than passive processing of information or accumulation of experiences, knowledge creation, diffusion, and transformation are actively and systematically conducted and embodied in all the organizational processes of production. It is these forms of learning that differentiate the capability of firms in the post-Fordist era.

Active learning entails doing so with purpose, preparedness and motivation. It is derived from people's intrinsic motivation, self-esteem, dignity, curiosity to learn, and joy in learning. Active learning involves systemic institutional arrangements that try to integrate conception with execution at all levels of an agency and synthesize existing knowledge and skills across departmental boundaries and along the value-chain of a product to create new knowledge. Such a process of integration and synthesis is generative, which is about creating, rather than adapting, which is about coping.

The Learning Organization

The natural result of focusing on active learning is the emergence of the learning organization. As one commentator has put it, '[It] is an organization skilled at creating, acquiring, and transferring knowledge, and at modifying its behavior to reflect new knowledge and insights' (Garvin, 1993). When thus defined learning is more than the sum of individual knowledge. The learning capability of an organization relies more on its institutional arrangement in facilitating interpersonal skills than on several bright employees who are excellent in learning.

The process of learning is the continuous process of experimenting, searching, articulating, and communicating to find opportunities to improve products and processes. To achieve this, one needs to continuously acquire firm-specific, tacit, relational, and localized knowledge through the process of learning. These kinds of knowledge are almost impossible to learn from textbooks. Firm-specific and tacit knowledge are usually acquired through learning by doing (Arrow, 1962), by using, by trying, through alliance, and by specially designed on-the-job training. Relation-specific knowledge is usually acquired through contacting and communicating with clients and users.

Sources of Learning

Sources of learning usually come from different levels of institutional arrangements that promote individual learning, organizational learning, transactional learning, network learning, spatial learning, and learning infrastructure. These have a different relevance for an agency such as the FAA.

Individual and Organized Learning

The learning capability of an organization depends on the knowledge and skill base of employees. In the Fordist age, individual learning relied heavily on formal training. This, however, only provides the knowledge base for further education; scientific knowledge acquired is only part of the whole body of skills and expertise needed for continuous improvement. The mastery of organization specific knowledge through learning by operating, using, and searching is sometimes more crucial than formal training. This transformation of tacit knowledge into explicit knowledge is demanding.

Effectiveness of individual learning in any organization depends on several things. First is the amount and quality of on-the-job training. Organizations that provide more effective training have an advantage. Second is employees' motivation to learn. Other conditions being equal, the greater the desire to learn the higher the learning capability. Gordon Bethune who engineered the turnaround of Continental Airlines in the mid-1990s, followed the maxim of placing incentives opposite the behavior that needed to be changed. Incentives motivate employees. The question needs to be raised: could the FAA have motivated employees? The fact that almost all of its projects ran over cost, and were seldom completed demonstrates the need for appropriate incentives.

The motivation for providing on-the-job learning depends on social and institutional factors – for example Japan, Taiwan, Singapore, and Hong Kong have cultural norms for learning. While such norms for learning are important, the nature of the employment relationship is a more influential factor. Short-term contractual relationships decrease employers' incentives to provide training, and reduce employees' motivation to learn organization-specific skills. There is also a loss of learning capability when an organization contracts and cuts its labor force in a cyclical financial downturn. Long-term employment relationships, however, motivate employers to provide training embodying organization-specific knowledge and skills.

The learning capability perspective institutionalizes the belief that improved quality also means reduced costs. The traditional system of mass production wasted the learning capacity and knowledge creation capability of shop floor employees. Moreover, its sharp division among functions with centralized control as the only legitimate coordination tool and over-emphasis on standardization without effective methods of continuously pushing forward and refining standards through shop floor learning also greatly limited organizations' capability in learning and knowledge creation.

The 'Japanese system' of manufacturing incorporates the best of the American system of mass production and the Swedish system of craft production. Delicate and in-depth knowledge-skills are developed through specialization. Through standardization, acquired knowledge and skills are further embodied and stored in organizational routines, and thereby overcome the bounded rationality of individual employees. Through the practice of total quality management, sources of dispersed knowledge are constantly elicited to improve operating routines, thus utilizing the learning and knowledge creation capability of individual employees.

As seen later, however, the FAA had no current capability in the early 1990s to improve operating standards since it focused only on completion or continuing projects. One exception was the Administration's improvement of noise standards.

Transactional Learning

Transactional learning happens in the 'market place' and is not limited to individual 'entrepreneurs.' More often than not, it includes organizations. A post-Fordist organization not only facilitates learning within its boundaries, it also takes every opportunity to learn from the marketplace, from suppliers to acquire the most advanced technology and related equipment and from customers to find new markets and to receive feedback on improving its products and services. What Kodama (1991) calls 'demand articulation' is a process of discovering the potential customer needs for a new generation of products.

The simplest form of transactional learning is when people are exposed to new ideas about products and ways of doing business in international markets. A second form is learning by using new equipment. New machines embody new knowledge, and the use of them facilitates a generation of new skills, expertise, and insights about opportunities for improvement.

With most of the energy at the FAA being deployed in the production process, all outside stimulation was lost. The FAA pushed microwave landing systems (MLS) in the 1980s and 1990s as its precision approach landing model despite the fact that new technologies had made the existing instrument landing systems (ILS) more efficient, and the potential for a global positioning satellite system (GPS) would seem to eliminate the need for MLS.

A third form of transactional learning occurs when people engage in an exchange of knowledge at specific times and places; or technological knowledge about specific products and processes, either formally or informally. A fourth form comes from the movement of knowledge workers across organizations. Experts in this category include those in engineering, marketing, and management. It permits knowledge to be exploited in divisions and units other than where it was developed.

Transactional learning would have greatly benefited the Administration, yet no process was in place allowing it to receive the precise signals needed from the 'marketplace' in which it operates. Even though many would have the FAA totally separate its oversight functions from any contact with industry, by doing so, those who are most in need of improved knowledge to undertake tasks such as airplane inspections, are cut off from those who benefit the most from this knowledge. While a captured regulator is not desired, especially in the area of maintenance inspections, there must be some interaction between the two groups in order to facilitate the exchange of knowledge and the improvement of skills. It is doubtful that a government agency or any private company on its own can improve performance in a vacuum of ideas.

Network Learning

Structurally, learning is enhanced by inter-organization networks in the market that facilitate knowledge flows among suppliers and producers. A prime factor in reforming the FAA was getting the organization to realize that it is a part of a very dynamic and constantly changing industry. If the airline industry worked at the same learning pace that characterizes the Administration, it would have bankrupted even faster than it did in the early 21st century. At some point when the airline industry is allowed to grow and change at the rate it needs the FAA must be brought into the learning process.

Culturally, various forms of social capital facilitate information exchange, especially the cultural norms of trust, cooperation, and reciprocity. Without these concentrated institutional structures and cultural norms, the market in the neo-classic sense can help little in facilitating and stimulating cooperative learning.

While transactional learning in the Fordist age was embedded in impersonal and often unreachable markets, transactional learning in the post-Fordist age is increasingly embedded in socially constructed networks. A post-Fordist organization tries to increase its learning capability by transcending unreachable market transactions of the Fordist age and to engage in the establishment of sustainable and effective cooperation by inter-organization networks and strategic alliances for learning and knowledge creation.

Whereas inter-organization relations in the Fordist age handicapped undertakings' abilities to learn from each other, sustainable inter-organization networks of complementary specialization in the post-Fordist age enable related organizations to access knowledge and competencies far beyond their bounded capabilities. They also make it possible to focus on core competencies and enhance their knowledge bases through the sharing of capabilities among networked organizations.

Underlying the need for trust and reciprocity is failure in the sharing of information. Ownership of information is not exclusive and therefore cannot be correctly priced. Before a buyer agrees to pay for ownership, the information must be accessed to evaluate its value. However, once received, the information is stored in an individual's brain and he/she could refuse to pay. In this case, the seller cannot erase information from the buyer's memory. One way to overcome this is to establish a long-term exchange relationship based on trust and reciprocity. If the organization believes that another will share its information reciprocally, an organization will probably share the information with the organization in advance.

Mutual benefits acquired through shared information are a realization of the potential positive externality of information. The same is true of complementary knowledge and capabilities within a network. The increase in core competencies in one organization enhances the learning capability of those networked through their cooperative sharing of knowledge and capabilities. Inter-organization networks can be seen as institutional arrangements that internalize the potential positive externality of information, knowledge and complementary competencies to promote learning capability. The existence of potential positive externalities, combined with the selection process, make possible the emergence of networks promoting the learning capability of all those involved.

However, effectiveness and sustainability necessitates a change in the nature of inter-organization games to a balance between competition and cooperation. It also requires the support of the cultural norms of trust and reciprocity as well as spatial, ethical, and communal linkages (Fukuyama, 1995).

Microwave Landing Systems

The Issues and the System

The purpose here is not to engage in a discussion of the pros and cons of the microwave landing systems (MLS). Rather it is to point out the organizational learning problems that caused the FAA to support its development of MLS for 13 years in opposition to those who would never use it.

The efforts to up-grade and modify landing systems was part of the much larger and longer running efforts of the FAA to keep pace with traffic growth and changing technology, while at the same time contain costs (Heppenheimer, 1995). The FAA was also seeking to adopt systems that enhanced air transportation safety as traffic flows were growing.

In 1981, the FAA initiated the National Airspace System Plan (later from 1990 the Capital Investment Plan). The justification for this was subsequently summed up by the FAA Administrator, Lynn Helms (1982):

> The present system does have serious limitations. It is labor intensive, which makes it expensive to operate and maintain. Even more important, it has very little ability to handle future traffic growth or future automation needs. These limitations result from an aging physical plant and inefficient procedures and practices. For instance, the present system still has many vacuum-tube systems.

The FAA's air traffic management system in the 1980s was thus seen as outdated and inefficient (Garrison, 1980). The Administration's plan to modernize initially covered a ten-year period and envisaged an expenditure of $12 billion but rapidly ballooned. By 1994 it was a $36 billion plan extending into the 21st century. The increased costs partly reflected the addition of new projects but also embodied an increasing number of cost-overruns.

Much of the effort was focused on on-route control and circumventing the limitations of radar that only has an effective range of 200 miles or so. Here there were problem almost from the outset. In particular, the FAA insisted on retaining programs for its major terminal radar approach control (TRACON) centers in the form of a proprietary code called Ultra, written in assembly language. The standard computer was the antiquated Univac 8303.

The problem was that the air route traffic control centers (ARTCCs) that interface with TRACON had much more sophisticated hardware and software. The result was periodic crashing of TRACON due to overload. Updating by using more powerful hardware for TRACON was not possible because more modern workstations could not accommodate Ultra and the FAA would not rewrite it. Instead a program was initiated to update the Univac hardware relying on a single supplier, Unisys.

In addition to radar, there was also a perceived need to up-grade landing systems. Microwave Landing Systems are part of what are called all-weather precision landing systems, an integral part of the nation's air traffic control system. They supply azimuth, elevation and distance information. They provide aircraft with precision guidance, enabling pilots to land safely on runways when low visibility conditions exist. These systems also enhance safety during good weather conditions. For an aircraft to use a precision approach, avionics equipment must be onboard to receive the electronic signal transmitted by the ground system. After receiving the signal, the onboard avionics provide an approach path for exact alignment and descent to a runway.

The ILS is the dominant precision approach and landing system used worldwide. Introduced in 1939, the ILS was adopted nationally in 1941 and accepted as the international standard in 1949. MLS provide all the functionality of ILS, with more precision and capability, and offers flexibility in choosing a landing approach. It has been widely used by the US military, and in locations where physical obstacles such as mountains prevented conventional ILS being used.

The limitations of the ILS were perceived before the creation of the National Airspace System Plan. But while there was pressure from the airlines to replace the ILS from the mid-1960s because of the vagaries of the system, and because with only 200 systems for the 2000 US airports congestion was increasing, progress was slow. The MLS technology was initially developed from the late 1960s by a number of companies each with a different system.

The Failure of the System

For the MLS to be adopted it had to meet the international standards imposed by the International Civil Aviation Organization (ICAO). Other countries, such as the UK and France, had preferred to deploy high-performance ILSs. It was not until April 1978, that the International Civil Aviation Organization determined that ILS would be replaced by MLS at international airports. The US Time Referenced Scanning Beam principle was the ICAO choice. The ICAO developed a transition plan in 1981 and embedded MLS in its Standards and Practices (Dehaynain, 2002). In 1986 the transition plan was revised to make 1986 the ILS-MLS turning point.

In 1983 the FAA awarded a contract for $79 million to Hazeltine – a firm with no previous experience in the field – to develop and set up 178 MLS airport systems. The FAA was to receive the first by 1985. The FAA, however, made numerous requests for technical changes that were both costly and caused delay. By 1989 Hazeltine had overcome many of the technical issues and had installed two units, although they still did not comply with FAA specifications. But the company requested an additional $100 million to complete the program. The FAA terminated the company's contract.

New contracts were awarded to Raytheon and Wilcox – suppliers that at the time of the award of the first contract had extensive experience of competing against European companies such as SEL and Phillips in the ILS market. But by this time the airlines were expressing strong support for the GPS navigation system that allowed pilots using simple receivers to determine their position and to land in bad weather with ease. In 1994 the FAA terminated the new contract. Canada and some other countries quickly followed the FAA's announcement by pulling out of the field.

The ICAO was also reconsidering its position at this time in the light of increasing interest in GPS systems, the retreat of the US, and the now increasing interest of some European states. A conference on the topic was held in 1995 and the pronouncement was that given the diversity of markets for air transportation, systems should be adapted to meet users' needs. Multi-mode receivers on planes would enable interoperability between airborne receivers and ground transmitters.

The cost of the pursuit of the MLS by the FAA was high. The US Department of Transportation (DOT) in an audit in 1987 had updated estimates of the cost to the government and users. At this stage, the Administration planned to spend about $1.5 billion to develop, acquire, and install MLS ground units. Estimated costs for users to acquire, install and utilize MLS avionics ranged from $500 million to $1.3 billion. Through fiscal year 1990, FAA spent approximately $500 million for research, development and program support.

Several points are noteworthy. The airlines had stated their opposition to this program from the mid-1980s and backed up their statements by not outfitting aircraft purchases with avionics that would interface with MLS. One of their initial concerns was the cost of MLS at a time when they were making heavy investments linked with new surveillance systems (e.g., TCAS) or communications systems (SATCOM). Despite the fact that no planes were being fitted for this technology, the FAA continued its development, possibly due to European rather than American interests.

Kenneth Mead, Director of Transportation Issues at the Government Accounting Office (GAO) commented at a hearing before the Subcommittee on Aviation of the Committee on Commerce, Science, and Transportation on August 2, 1995 that

> Probably every Member of Congress that has been around for the last 12, 13 years can remember the microwave landing system. Every year it would crop up. The airlines did not seem to want it, but FAA kept pushing it. Finally, FAA said, we see satellites as the way to go, and people said, 'finally,' and FAA has been moving out smartly ever since, but it seemed like it took a long time. The MLS program was a program with nine lives.

In light of current learning organization research, it is interesting that Mead used this statement to support the contention that,

> [The] FAA over the last two years, is coming to grips with, grabbing by the neck, you might say, some problems that have been with us for a long time.

In response to Mead's statement, one Senator replied,

> Well, what you are saying to the committee is in the last couple of years they have been grabbing things by the nape of the neck and getting things done. Making pretty good judgments and moving us into the future, and I can see all three of you shaking your heads, so that makes me feel pretty good.

What is of most interest is Mead's statement that it took a long time for the FAA to move away from a bad decision, even when the users were opposed to it.

Rather, the way of thinking is indicative of a learning dysfunctional agency. This is a structural problem. The issue is why the Administration took so long to make the decision when there was so much opposition to the MLS strategy? It is to be expected that after thirteen years of being told the system was not wanted by the users, that it would have been abandoned. Why did it take thirteen years, despite opposition from the people who would use it, to finally abandon the system?

The initial logic was correct. According to the FAA, ILS systems had certain limitations. First, it was site sensitive and expensive to install. Because the ground in front of the glide slope antenna is used to form the beam, a large area in front of the antenna must be level. The cost to prepare sites for this sometimes exceeds the cost of the equipment.

Second is glide slope sensitivity to nearby reflecting surfaces. This can reduce airport capacity where departing aircraft must be held at a long distance from the takeoff threshold at certain airports in order to stay clear of the critical reflections area when an instrument approach is in progress.

A third limitation involves the 40 ILS channels available and the problem of frequency congestion in several parts of the country. It is already difficult to add ILS facilities to congested areas such as Los Angeles and New York.

Fourth is that ILS lacks the flexibility needed for future aircraft operations. The system provides only a single glide path and is not adaptable to high-angle approaches by short take-off and landing vehicles or rotocraft. It provides azimuth guidance to a single approach path over a very narrow sector.

Lastly, signal reflections limit ILS applications in areas of rough terrain, mountains, and large man-made structures.

At the same time the FAA was releasing figures on MLS and promoting its virtues, the Air Transport Association (ATA), that represents the bulk of the carriers in Washington on technical matters, began releasing its own assessment on MLS. The ATA had initially advocated in the late 1960s and 1970s some form of landing system relying on microwaves, but had subsequently become disillusioned with the MLS that was emerging. Its analysis by the 1980s differed greatly from that of the FAA. The ATA recommended that ILS should continue as the primary precision guidance system.

The DOT's 1987 audit concluded

> Our review showed that one of these systems, the MLS, has not been fully accepted by the aviation community. Consequently, our audit concentrated on whether the intended beneficiaries of the MLS plan to equip their aircraft with the necessary avionics in sufficient quantities to justify FAA's plans to purchase and install 1,250 MLS ground systems at a cost of $1.5 billion.

In response to this lack of enthusiasm, the DOT concluded that the '...FAA plans to replace the existing ILS with MLS even though the principal beneficiaries of MLS are not currently planning to equip their aircraft with the necessary avionics to utilize the system.' Even though the airlines were not going to use the system, the FAA was continuing with development plans. The airlines argued that during the time period the Administration was trying to develop MLS, ILS had improved. Further, since airports were not gaining capacity at any level, there would be no benefits to changing systems.

The story did not actually end in 1994. The FAA requested funds in fiscal years 1997 and 1998 to proceed with procurement of 500 additional MLS ground systems. The basis for this being that advances in GPS had been more problematic than anticipated. Congress did not approve any funds for MLS in financial year 1997 and tentatively zeroed out funding in financial year 1998. FAA requested $98.6 million for procurement of MLS ground systems in financial year 1999 but the program was dead.

This persistence of the FAA to develop a system not wanted is a testament of its Fordist-production orientation. The lack of a reasonable network sensitive to the factors in the market causes the FAA to always be a determinant to the industry it oversees.

Conclusions

In applying post-Fordist learning capacity methodology to the FAA, it is important to identify those who create the various factors promoting transactional learning and how they interact. Yet the Administration in its own documentation established that it considered one of its prime customers to be the air traffic controllers. Although the air traffic controllers are vital, the main beneficiaries are the airlines and the traveling public. It is these groups to whom the FAA should have been paying attention. Post-Fordism implies that relationships with all groups, including the controllers, is not only necessary but critical. Those who benefit should have been interacting with the Administration, and the FAA needed to establish strong reciprocal links.

The absence of these relationships eliminated any possibility of transactional learning. How these relationships are managed will determine how well the FAA's operation performs in the future. A lack of ability to learn from any marketplace was its greatest weakness. One of the corollaries to the absence of transactional learning is the concept that some types of knowledge do not come from books alone. Interaction is the learning process. It is an inescapable conclusion that the FAA did not understand the industry it was regulating without interaction. No matter what the FAA did, in the absence of transactional learning it always made the wrong move.

Because no clearly articulated definition of the main beneficiaries existed at the FAA, there was no accountability. This created problems for the airlines interacting with the Administration that did not know how its operations and policies affected airline operations. Implementing some type of cost accounting system could have helped. The FAA had no idea how it was affecting the industry since it did not know the costs of moving a plane through the system. A second step would have been the establishment of user fees to bring the FAA and the airlines into a closer working relationship. A user fee serves as an incentive to the airlines to work more closely with the FAA and to realize how much the Administration affects their operations.

The closure of ValuJet demonstrated to the public just how deprived the FAA was of knowledge of the industry it regulates. It took inspectors over four years to determine that there was a problem with the airline. It should not take any organization this much effort to function. The FAA's lack of accountability made it difficult for the organization to react or learn. For the Administration to acquire new knowledge, tremendous force from the outside was applied. In took a major airline crash to expose the deficiencies within the system.

Acknowledgement

The authors would like to acknowledge the input of Dengjian Jin in the preparation of the theoretical under-pinning of this chapter.

References

Dehaynain, C. (2002) *MLS, 10 Years Later...*, *ATNA Technical Review*, 62, pp.103.

Fukuyama , F. (1995) *Trust: The Social Virtues and the Creation of Prosperity*, The Free Press, New York.

Garrison, P. (1980) *How the Air Traffic Control System Works*, Tab Books, Blue Ridge Summit.

Garvin, D, A. (1993) 'Building a Learning Organization,' *Harvard Business Review*, vol.71, July/August, pp.78-91.

Helms, J.L. (1982) *Astronautics and Aeronautics*, June, p.51.

Heppenheimer, T.A. (1995) *Turbulent Skies: The History of Commercial Aviation*, Wiley and Sons, New York.

Kodama, F. (1991) *Analyzing Japanese High Technologies: The Techno-Paradigm Shift*, Pinter Publishers, New York.
Porter, M. (1990) *The Competitive Advantage of Nations*, Free Press, New York.

Index